Gas Pipeline Renewal
Insertion Technology

Gas Pipeline Renewal
Insertion Technology

Hayat Ahmad

Gulf Publishing Company
Houston, London, Paris, Zurich, Tokyo

Gas Pipeline Renewal
Insertion Technology

Portions of this book first appeared in *Pipe Line Industry* magazine, a Gulf Publishing Company publication.

Library of Congress Cataloging-in-Publication Data

Ahmad, Hayat.
   Gas pipeline renewal: insertion technology/Hayat Ahmad.
      p. cm.
   ISBN 0-87201-308-1
   1. Gas, Natural—Pipe lines—Maintenance and repair.   I. Title. TN880.5.A27   1990
665.7′44—dc20   89-48335
                                                        CIP

Contents

PROBLEM ASSESSMENT

PROJECT MANAGEMENT

PROCEDURES

Foreword

Gas Pipeline Renewal is an expanded version of a series of well-received articles that originally appeared in *Pipe Line Industry* magazine. Renewing older gas distribution systems safely and economically is becoming more important to the owners and operators of aging gas distribution systems.

Owners of these older gas distribution systems all over the world are searching for the latest techniques, materials, and engineering practices to extend the useful lives of underground facilities that can be more than 75 years old. Many of the aging lines are cast iron, mechanically coupled, welded steel, or wrought iron and ductile iron. At the time they were installed, the material was the latest available.

Now, many years later, these older buried facilities still do the job for which they were built, but the cost of upkeep and maintenance keeps increasing (just like for us aging humans). In addition, many of the older systems in Europe, South America, the Middle East, and Asia were designed for and installed to transport low-pressure manufactured gas. Now that natural gas is available and the systems are being converted to this energy source, the older pipeline facilities are really not satisfactory.

The number of these conversion projects is increasing, and owners are searching for the most effective engineering practices, installation techniques, and new materials—just exactly what Hayat Ahmad describes in this book.

Mr. Ahmad has worked in many phases of the gas distribution industry since 1967. His training and experience make him an authority on gas utility subjects. *Pipe Line Industry*'s readers found his series of articles not only timely, with the most recent advances in technology and materials described, but also filled with useful hints on how to renew older gas distribution facilities safely and at a cost that an owning company can afford.

William Quarles
Editor, *Pipe Line Industry*

Preface

This book deals with replacing gas pipelines by inserting new pipes into old lines. It is intended particularly for natural gas companies, and is based primarily on work done from 1986 through 1988 at Northwestern Utilities Limited, located in Edmonton, Alberta, Canada. Some of the analysis, therefore, is based on gas utility construction costs in Alberta in 1988. However, the principles given here are universal, and applicable to any other type of pipelines.

The reader can gain a general understanding of insertion technology from this book, and apply that knowledge to his specific needs. The assumptions and calculations made, however, should be checked against prevailing circumstances before proceeding with any pipeline replacement job. In certain countries, patent rights may be in force on live insertion, for which a fee may have to be paid to the patentee.

Currently, neither metric nor English units are consistently used in the North American gas industry. Although much of the design and legal work is in metric units, the man in the ditch speaks what he finds convenient. The lack of availability of metric pipes makes it more difficult to give up English units. As a result, the gas industry is still using both the OD (mm) and nominal pipe size (NPS) designations. This book reflects both where possible, mirroring the usage of the industry.

It is hoped that this book will provide useful information to pipeline companies, particularly gas utilities, whose pipes may need replacement.

Hayat Ahmad
Edmonton, Alberta

Acknowledgments

I would like to thank all those who assisted me in the preparation of this book. Much of the work contained here is based on project work carried out by the author at Northwestern Utilities Limited of Edmonton. Particular appreciation is due to Northwestern Utilities Limited for allowing me to write about the work performed there, and to D. H. Bailey, D. L. Oatway, and Ben Sokol for reviewing chapters of this book. My thanks also to Janice Froese and Mavis Jankowski for many hours of painstaking typing. I look forward to hearing any comments about this book from my readers.

1
Economic and Risk Analyses

INTRODUCTION

An airtight pipeline system requiring no repairs is an ideal system for which to strive, as it is not only wasteful but also expensive to lose valuable pipeline products. Pipeline repair costs are ever on the increase. Gas leaks should not be tolerated, nor should high-cost maintenance escape scrutiny. With the deterioration of the components of the pipeline, such as the pipe, fittings, and joints, a point is reached when the replacement of the original pipeline becomes necessary, either due to a lack of safety and reliability, or due to the economics, or both.

Before a gas utility embarks on a wholesale pipe-replacement project to replace aging cast-iron or steel pipe, it is important to collect data on the history of the repair frequency, and the costs of previous repairs for the pipeline system that is being replaced. These data should go as far back as possible—at least five years and preferably ten years should be studied. An evaluation for the system replacement should then be carried out and recorded. Such a record may be required in subsequent rate hearings before the Public Utility Board.

Data required for the risk analysis are far greater than those required for the economic analysis for which only the leak and repair history and costs of repairs and replacements are required. For the risk analysis, a number of other parameters such as the age, diameter, coating, and pressure must be known. These parameters and a suggested scoring system are explained later under the heading of "Demerit Point System." The analysis given under the demerit point system is a very simple and practical way of analyzing risk for gas pipe. However, the reader is cautioned that it is not a comprehensive risk analysis in the true definition of the term "risk analysis," which deals vigorously with the statistical data and the theory of probability. The reader is advised to refer to a textbook on risk analysis for better understanding of this subject.

DATA COLLECTION

The most convenient way to collect data will vary from company to company, as it is dependent upon the form in which the data were first recorded. For an urban gas utility, a repair history on a block-by-block, or a whole-street basis may be available. For a rural pipeline system, a repair record on a kilometer basis or township basis may have been kept. No matter in what form this record is kept, it must be changed into units of length consistent with the geography of the area and the age group, type of material, size, pressure, class location, etc. In some cases a block-by-block analysis may be appropriate, while in others a block may be too small a unit and have an insufficient number of data points, thus giving the data a low statistical validity. A convenient unit of length for data collection is a 3-city-block length or $1/2$ km (or $1/3$ mile).

Repairs that were necessary due to corrosion, faulty material or workmanship, or breaks or leaking joints are the only type that should be considered. Incidents of external damage, valve leakages, etc., must be discounted as they do not represent the deterioration of the pipeline.

Additionally, work-order files, service-order files, construction drawings, foreman's reports, and so forth will need to be scrutinized to collect data required for the risk analysis, as detailed in the demerit point system.

ECONOMIC ANALYSIS

An economic analysis should consider the incidence of repair in the past, project repairs for the future, and estimate the repair cost that will be incurred if the pipe is not replaced. It should also estimate the current value of the pipe and depreciate it for future years. Graphs can be drawn showing the rising number of repairs, the rising cost of maintenance, and the declining value (assuming an appropriate depreciation rate). Replacement should take place prior to or at the break-even point; in other words, the year in which the estimated maintenance cost equals the depreciated value.

Figure 1-1 shows an increasing number of leaks. Leak-report data for several pipes of similar characteristics (such as steel material, bare, or unprotected), for a unit length of 0.5 km ($1/3$ mile) or a similar unit chosen as per the considerations given under "Data Collection" is used to develop an equation for the number of leaks per year, N. The number of leaks in any old pipeline system tends to increase exponentially. Exponential growth of leaks will fit the following equation:

$$N = N_1 e^{nA}$$

where N = number of leaks for the study area in any year considered
         N_1 = number of leaks in the first year for which leak data are available
         e   = 2.718
         n   = nth year for which the number of leaks is to be estimated
         A   = leak growth-rate coefficient

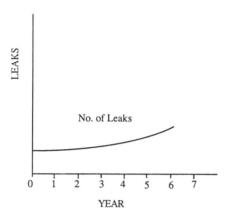

Figure 1-1. Typical leak history before replacement.

The coefficient A can be calculated from the graph drawn of the leak data for the study period. The study period should preferably be in the range of 5 to 10 years. Any data less than this will not give enough points to accurately determine the coefficient A. Before a graph is drawn, leak report quantities must be adjusted for any changes to the system during the study period. In other words, if a township was the study area, then pipe replaced, added, abandoned, or installed should be tallied. If this is not done, errors will result. For example, if leak data were collected on a township basis, and during the study period considered, 10% of the township's pipe was renewed, leak frequency in the old area will continue to rise while in the renewed area there should be none, possibly giving an overall number of leaks less than that of the year prior to the replacement. For an accurate picture and a determination of the leak growth-rate coefficient, the number of leaks in the years following the replacement should be accounted for as follows:

Let x = percentage of replacement completed

Let y = actual number of leaks for remainder of the township for the year considered

Adjusted number $N = \dfrac{y}{1 - x}$

The leak growth rate for some pipelines may not follow the exponential curve. For example, in a particular case, a linear increase with time may be predicted. In that case, use a linear equation instead of the exponential form. Regardless of the specific form of the leak forecast equation, the general features of the economic analysis given in this chapter will apply.

Example 1-1

From the best-fit curve of an actual leak history of a gas utility, the following numbers are noted:

N_1 = number of leaks in first year (i.e., 1979) = 140
N = number of leaks in year considered (1986) = 310
$n = 86 - 79 = 7$
$310 = 140 \ e^{7A}$

Hence, coefficient $A = 0.113$. The estimated number of leaks in 1990 will be:

$N = 140 \ e^{11 \times 0.113} = 485$ leaks

Figure 1-2 shows the typical unit repair cost in constant dollars. With a piping system that is corroding, not only do the number of repair visits increase, but also the repairs required become larger and larger as the system gradually worsens from patch repairs to cut outs. Annual repair costs may vary significantly from year to year, hence choose the best-fit curve. Remember to convert all costs in constant dollars.

Figure 1-3 shows the typical maintenance-cost curve (constant dollars). This curve is obtained by multiplying the results of Figure 1-1 and Figure 1-2. Figure 1-4 shows the rising maintenance-cost curve and the depreciated-value curve. The intersection of the two curves determines the year in which the replacement of the section of pipe should occur, as beyond that point the company will be spending more money on repairs than the pipe value. No doubt the same technique can be applied for the replacement of any other equipment, be it underground, aboveground, or mobile.

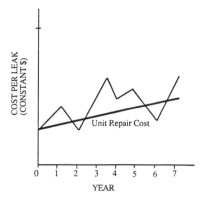

Figure 1-2. Typical cost per leak (constant dollar). Not only does the number of leaks increase over time, but the extent of needed repairs increases as well, i.e., progressing from a simple patch repair to a cylindrical section replacement.

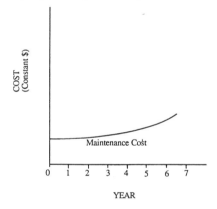

Figure 1-3. Typical maintenance cost.

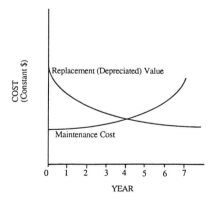

Figure 1-4. Typical maintenance and replacement value.

RISK ANALYSIS: DEMERIT POINT SYSTEM

In the demerit point system, emphasis is placed on safety rather than on economics to determine which pipeline must be replaced. This method is statistical and requires a great deal more data. It is based on two elements:

1. Leak potential of pipeline.
2. Potential hazards if a leak occurs.

Arbitrary values based on the judgment of the operating staff may be given to the various physical properties of the pipeline. Yet another arbitrary value is selected to pass or fail the pipe. Pipelines scoring more than 50% are chosen for replacement while those scoring less can stay in service until disqualified at a later time.

A pipeline with the highest leak potential and the highest potential risk is given the highest score, whereas a pipeline with the least leak potential and risk is given a zero score. The closer a pipeline is to the highest possible score, the greater the need to replace it. The company may allocate a certain amount of its budget for its pipe-renewal program. Within the given budget, certain pipelines scoring high on the demerit point system may be chosen each year for replacement.

The demerit point system is an aid in determining the urgency for renewal, but it should be used with due care. Considerations such as the street-repaving program, load growth, intermixing of steel/cast iron, and PE (polyethylene) pipe material, pressure problems, and cathodic protection requirements may necessitate the lower-scoring pipe to supersede the pipe with the higher score.

Following is typical demerit point system for gas distribution steel mains. A company may evolve its own demerit system to suit the conditions, the pipe material, and the preference of its operators.

Score	Remarks
0–25	Generally acceptable
26–40	Surveillance necessary
41–50	Consider replacement
51–65	Replacement program item
Above 65	Unsafe—shutdown recommended

This may be depicted by a color chart as follows:

Acceptable	Surveillance	Alert	Replace	Danger
Green	Gray	Purple	Pink	Red

0   25   40   50   65   100

Demerit Points for Steel Mains

Leak Potential   **Points**

1. Age
   a. Newer than 10 years   0
   b. 11–20 years   5
   c. 21–30 years   7
   d. 31–40 years   9
   e. 41–50 years   11
   f. 51–60 years   13
   g. Older than 60 years   15
2. Wall Thickness
   a. > 10 mm (0.0390 in.)   0
   b. 6 mm to < 10 mm (0.236 in.–0.390 in.)   1
   c. 4 mm to < 6 mm (0.157 in.–0.232 in.)   2
   d. 2 mm to < 4 mm (0.079 in.–0.157 in.)   3
   e. < 2 mm (0.079 in.)   5
3. Coating
   a. Yellow jacket or epoxy (or paint on exposed pipe)   0
   b. Tape, asphalt, or enamel (buried pipe only)   4
   c. Bare (buried or exposed pipe)   8
4. Cathodic Protection
   a. Rectifier   0
   b. Anode   1
   c. None   9
5. Joints
   a. Welded   0
   b. Mechanical   4

6. Soil Resistivity (ohm·cm)
   a. Over 10,000 or for aboveground pipe   0
   b. 5,001–10,000   1
   c. 2,001–5,000   2
   d. 1,001–2,000   3
   e.   501–1,000   4
7. Leaks Due to Material Faults or Corrosion
   in Last 10 Years per 0.5 km (1/3 mile)
   a. < 1 leak   0
   b. 1 leak   4
   c. 2 leaks   8
   d. 3–4 leaks   16
   e. > 4 leaks   20
8. Observed Condition of Pipe
   (e.g., corrosion pits, loose coating, etc.)
   a. None   0
   b. Some (up to 2% surface area)   2
   c. Severe (over 2% surface area)   5

Risk Potential

9. Class Location
   a. Class 1   0
   b. Class 2   2
   c. Class 3   4
   d. Class 4   8
10. Crossings
    a. No crossing   0
    b. River or other water body   1
    c. Railway tracks   2
    d. Highway   3
    e. Bridge, railway, or automobile   6
11. Operating Pressure
    a. 0–2 kPa (0–8 in. w.c.) (LP)   0
    b. > 2–105 kPa (8 in. w.c.–15 psi) (MP)   2
    c. 106–210 kPa (15–30 psi) (IP)   3
    d. 211–410 kPa (30–60 psi) (IP)   4
    e. 411–550 kPa (60–80 psi) (IP)   5
    f. 551–3,500 kPa (80–700 psi) (HP)   7
    g. > 3,500 kPa (> 700 psi) (HP)   8
12. Diameter
    a. ≤ 60 mm (2 in.)   0
    b. 88 mm to 114 mm (3 in.–4 in.)   4
    c. 168 mm to 323 mm (6 in.–12 in.)   6

13. Cover and Loading
    a. Statutory cover and loading per design   0
    b. Inadequate cover   1
    c. Excessive loading   2
       (over the design limit for the depth of
       cover)

Scoring

Best possible score   0
Worst possible score   100

Example 1-2

114-mm (4 in.) diameter, 3.2-mm (0.125 in.) wall, low pressure, steel main, laid in 1924, no coating, no cathodic protection, method of joining pipe was Dresser couplings, buried in ground having a soil resistivity of 900 ohm·cm at adequate cover under city street pavement in class 3 residential area with light traffic, has a history of 2 leaks in a 6-block (1 km or ⅔ mile) length during the past 10 years. Observed condition of pipe elsewhere to be good.

Particulars	Points
1g	15
2d	3
3c	8
4c	9
5b	4
6e	4
7b	4
8a	0
9c	4
10a	0
11a	0
12a	0
13a	0
Total Score	51

Recommendation: This main should become part of the replacement program.

Example 1-3

30-year-old IP (550 kPa or 80 psi) steel main, 5.6-mm (0.22-in.) w.t., 323-mm (12 in.) diameter, insulated from buried pipeline system, painted, crossing a ¹/₂-km (1,640 ft) wide river bridge in class 3 area; outside pipe condition where visible found to be good, but in the last 10 years, had two repairs done on the bridge, one due to the partial collapse of the bridge and the other a pin hole at a welded joint. The valves on this section also leaked and were repacked twice in the last 10 years.

Particulars	Points
1c	7
2c	2
3a	0
4c	9
5a	0
6a	0
7b	4
8a	0
9c	4
10e	6
11e	5
12c	6
13a	0
Total Score	43

Note: Ignore irrelevant data.

Recommendation: The demerit point system recommends that pipelines scoring above 40 be considered for replacement and those scoring above 50 be placed in the replacement program. Therefore, the condition of this line should be investigated further. A pressure test or non-destructive testing may reveal further pipe material defects which may increase the score to above 50, putting it into the replacement program.

Demerit Points for Iron Mains

Most old gas systems throughout the world have cast-iron mains with mechanical or lead joints. With the advent of ductile iron in the second half of this century, many gas companies switched partially or completely to ductile iron. The ductile-iron pipe is stronger and more ductile than the cast iron, in which the graphite is present in flake forms. Although there are some six known types of pipe-joint configurations for the iron pipes, they can all be classified into two general types: a mechanical type in which a resilient gasket under compression seals the gas, and a rigid type in which the seal is made by caulking soft metal between the bell and the spigot. The ductile-iron pipes are assembled using a mechanical joint rather than the lead/jute combination. The mechanical joint is considered superior to the lead joint as it allows a certain degree of flexibility to pipe movement. However, prior joint cleaning and proper bolt tightening must be done with care if leak-proof joints are to be achieved. Most of the old gas systems carried manufactured (town) gas. When natural gas became available, these systems were converted to natural gas. While natural gas is superior to manufactured gas in many ways because it has a higher calorific value, negligible carbon monoxide content, and is drier, it did adversely affect the cast-iron pipes, as it tended to dry out the lead/jute joints. Also, the absence of tar in natural gas further exposed the seals at the lead joints. Natural gas may also contain trace amounts of hydrogen sulfide gas (H_2S), which adversely affects cast and ductile irons.

Many of these older systems are near the end of their useful life. Cast-iron and ductile-iron pipes were buried bare, joined together by gasket-type joints or by pouring molten lead between the bell and spigot of the pipe ends. Cathodic protection was not provided. It would have been fairly difficult and expensive to provide cathodic protection to iron pipes, which lacked coating protection and, in many instances, the electrical continuity.

Following is the demerit point system for iron mains:

Leak Potential	Points
1. Age	
a. Up to 10 years	0
b. 11–20 years	5
c. 21–30 years	7
d. 31–40 years	9
e. 41–50 years	11
f. 51–60 years	13
g. Older than 60 years	15
2. Wall Thickness	
a. Class 23	0
b. Class 22	1
3. Restraints on Bends	
a. Restrained properly by thrust block, etc.	0
b. Not restrained	6

4. Soil Stability
   a. Stable soil   0
   b. Unstable soil or frost penetration up to the pipe   9
5. Joints
   a. Bell and spigot joint with resilient gasket or
      other mechanical joints   0
   b. Lead/hemp joints   4
6. Soil Resistivity (ohm·cm)
   a. > 10,000 or for aboveground pipe   0
   b. 5,001–10,000   1
   c. 2,001–5,000   2
   d. 1,001–2,000   3
   e.   501–1,000   4
7. Leaks Due to Material Faults or Corrosion in Last
   10 Years per 0.5 km (1/3 mile)
   a. < 1 leak   0
   b. 1 leak   4
   c. 2 leaks   8
   d. 3–4 leaks   16
   e. > 4 leaks   20
8. Observed Condition of Pipe
   (e.g., corrosion pits, graphitization, etc.)
   a. None   0
   b. Some (up to 2% surface area)   2
   c. Severe (over 2% surface area)   5

Risk Potential

9. Class Location
   a. Class 1   0
   b. Class 2   2
   c. Class 3   4
   d. Class 4   8
10. Crossings
    a. No crossing   0
    b. River or other water body   1
    c. Railway tracks   2
    d. Highway   3
    e. Bridge, railway, or automobile   6
11. Operating Pressure
    a. 0–2 kPa (0–8 in. w.c.)   0
    b. 2–14 kPa (8 in. w.c.–2 psi)   3
    c. 14–170 kPa (2–25 psi)   5
    d. > 170 kPa   7

12. Diameter
    a. ≤ NPS 4   0
    b. NPS 6–NPS 12   4
    c. NPS 16–NPS 24   6
    d. NPS 30–NPS 48   8
13. Cover
    a. 0.75 m to 2.5 m (2.5 ft–8 ft)   0
    b. > 2.5 m (8 ft)   2
    c. < 0.75 m (2.5 ft) (not permissible)
14. Gas Quality
    a. Natural gas   0
    b. Town gas   2
    c. Gas containing up to 1 mg/m^3 H_2S 3
    d. Gas containing 1 to 2.3 mg/m^3 H_2S 5

Example 1-4

NPS 4 diameter, class 22 wall, 14 kPa, cast iron main laid in 1924, no coating, no cathodic protection, method of joining pipe was lead joints buried in ground having a soil resistivity of 900 ohm·cm at 1 m cover under city street pavement in class 3 residential area with light traffic, has a history of 12 leaks in a 6-block (1 km or 2/3 mile) length during the past 10 years. Observed condition of pipe elsewhere to be good. Pipe bends were restrained when constructed. Soil is stable with no frost and the gas distributed is clean natural gas.

Particulars	Points
1g	15
2b	1
3c	0
4a	0
5b	4
6e	4
7e	20
8a	0
9c	4
10a	0
11b	3
12a	0
13a	0
14a	0
Total Score	51

Recommendation: This main should become part of the replacement program.

COMBINED ECONOMIC/RISK ANALYSIS

Utility companies are required by legislation to maintain a high level of safety in their operation. It is obvious, therefore, that economics should not be the only criterion for the replacement decision. Public safety is foremost in the minds of both the regulatory bodies and the gas companies.

The criteria given under the demerit point system may be chosen to first select those pipelines for which replacement is deemed necessary. Once these are sorted out, an economic analysis may be done to set the priorities and the sequence of replacement. At this juncture, it may be necessary to join consecutive sections of risky pipe. This is necessary for the following reasons:

1. Operational considerations and construction practices usually dictate that replacement work be carried out in consecutive steps rather than by moving crews all over the city.
2. Combined sectional or system figures for the number of leaks, maintenance cost, and replacement cost will provide better analysis, as a single section of $1/2$-km ($1/3$-mile) length may not provide sufficient data to deduce sensible results.

APPLICATIONS

The formulae and the methods given are simple and can be applied to any above or belowground distribution steel or iron pipeline, and with minor adjustments to PE piping systems. For a steel gas-transmission line, other factors that affect the risk potential and thus the demerit points include the remote sensing and controls availability (SCADA), the grade of the pipe, and the travel time from the operation base (crew availability for repairs). These items have to be worked into the demerit points when considering a transmission line.

This chapter has shown how to evaluate the conditions of plants in numeric forms and how to plan ahead and take corrective action well in advance. The recommended steps will ensure a higher degree of safety, reliability, and profitability.

2
A Case History

Increasing reports of leaks in a gas utility "Y" (GUY) service area worried the operations department. GUY had first started to install steel pipe in 1923, and by 1935 much of the city's commercial and adjoining residential areas were serviced by the bare steel pipes, joined by mechanical couplings. Although in GUY's service area winters are cold, with frost penetrating several feet below the gas mains, GUY did not suffer very much from frost heaving of the mains and subsequent mechanical couplings' pull outs, which had been a common occurrence in some other gas companies. However, GUY strapped the mechanical couplings whenever it excavated near a coupling. The use of the bare steel pipe continued from 1923 to 1934. In 1934, coal-tar wrapping was introduced, followed by wrapped enamel, which was used until 1964 when yellow polyethylene coating was introduced. Until this time, all bare pipe and most of the coated pipe remained without any cathodic protection, and it stood up very well against the environment. When cathodic protection was introduced, it was applied to the new yellow-jacketed pipe and to other older wrapped pipe that had welded joints. Cathodic protection (CP) on bare pipe would have been ineffective since it was joined together by insulating mechanical couplings.

The lack of coating and the cathodic protection allowed the normal course of corrosion to take place. During the late seventies and the early eighties, incidents of leaks significantly increased. Whenever a repair was made, the condition of the adjoining main was noted. This information led the operations department to conclude that some drastic measures were necessary.

In 1985, a study on corrosion control was done, which indicated that GUY was experiencing much higher leak rates than most other utilities. The basic cause of this experience was the low priority that was previously given to the corrosion control program. At this time, GUY was spending about

15

$1.24 million annually on leak repairs, or $3,400 per day, with a leak rate of 5.1/100 km (3.17/100 miles). The average leak repair for the company-wide operation was $2,285 per leak while for the central district with its bare mains, it was $3,313 per leak. Other gas utilities surveyed at the time had the following leak rates:

Company	Leaks per 100 kms	Leaks per 100 miles
1	0.08	0.05
2	0.48	0.30
3	0.10	0.68
4	2.15	1.34
5	3.75	2.33
6	5.10	3.17
7	9.19	5.71
Average	3.12	1.94

At this juncture, after reviewing the repair costs, the number of leaks, and the potential hazards, the management decided to:

1. Put more resources to cathodic protection with the aim of reducing the leak incidence to less than 1/100 km (0.62/100 mile).
2. Study the bare mains area as a special project.

The following chapters deal specifically with the Bare Mains Project. This is also referred to as the Urban Replacement Project or the Gas Distribution Renewal Project.

SCOPE OF PROJECT

Having come to the conclusion that the central district's bare steel mains and 50- to 60-year-old services were adding significantly to the company's leak statistics and repair costs, it was no surprise that the bare mains area was chosen for detailed study to:

1. Provide a detailed scope of the project.
2. Define the extent and magnitude of the problem.
3. Conduct an industry-wide survey to determine what others in similar circumstances had done.
4. Determine the replacement alternatives.
5. Provide a cost-benefit analysis of each alternative.
6. Recommend a strategy, including the replacement method.
7. Formulate a construction contract that could be tendered.

It was quite obvious that the uncoated (bare) and unprotected (having no cathodic protection) steel pipe was adversely affecting the leak statistics. We had to find how many bare mains and bare services there were. Plan sheets were reviewed, service order files were taken out, and regulating stations records were scrutinized. All bare mains and regulating stations in the bare mains area were plotted on large wall maps, approximately 2.5 m × 2.5 m. A scale of 1:5,000 was chosen for good legibility. Overlays were used to separate different sets of information, such as the low pressure, the medium pressure, and the intermediate pressure systems. The final count of quantities was as follows:

Mains

Size (mm)	Size (NPS)*	Length (m)	Length (miles)
42.2	1¼	2,088	1.30
60.3	2	10,814	6.72
88.9	3	25,578	15.90
114.3	4	82,203	51.09
168.3	6	33,194	20.63
219.1	8	19,154	11.90
273.1	10	17,091	10.62
323.9	12	9,662	6.00
Total mains		200,000 m[†]	124 miles[†]

[†]Totals rounded off
* NPS = normal pipe size

Number of Services

Low pressure (LP)	9,720
Medium pressure (MP)	8,453
Intermediate pressure (IP)	50
Total	18,223

Regulating Stations

Regulating stations	45
Regulating boxes (small reg. stns.)	23
Total	68

Priority Areas

Having determined the affected area and the number of mains, services, and regulating stations that needed to be replaced, revamped, or abandoned, the next step was to ascertain the sequence and the method of replacement.

Sequencing was a relatively easy task. Higher priority was given to areas of higher need. Areas with the highest number of leaks and the highest risk potential were given priority. In other words, a risk analysis was done.

CHOOSING THE REPLACEMENT METHOD

The direct burial method, dead insertion, and live insertion techniques were reviewed. The management considered the following options for the bare mains:

1. **Provision of Cathodic Protection To Bare Pipe.** This would entail locating all insulated mechanical couplings, bonding across them, and providing cathodic protection (CP) to the pipe. A massive quantity of cathodic current was estimated. Interference with other utilities and the subway system was also a consideration. The evaluation showed that, in spite of the high cost and a major CP effort, this alternative would not reduce the leak rate to an acceptable level, as the condition of the steel pipe was beyond recovery.

2. **Coating and Wrapping Bare Pipe.** Costs were high, with little benefit.

3. **Direct Burial.** The direct burial method makes the replacement task easy. Customers are turned off only for a brief period, when their individual services are switched over to the new main. This method also provides the flexibility to choose larger diameter pipes than the old ones. Similarly, it gives the freedom to select either steel (or cast iron) or PE pipe material. Note, a metal pipe inside another metal pipe is not a good idea as shorts occur, leaving the inside carrier pipe without cathodic protection. The direct burial method is much more costly than the insertion method. The direct burial method requires excavating a trench all along the pipe route. As it was so much more costly than the insertion method, it was decided that the direct burial method should be used only in those circumstances where insertion cannot be done. (For further discussion on this topic, the reader is advised to refer to Chapter 12, Criteria For Replacement.)

4. **Live and Dead Insertion.** In the insertion technique, new plastic pipe is placed inside the existing pipe, which then becomes redundant. The

gas is carried by the new plastic pipe, and the old metallic pipe (which now contains no pressure) acts as a casing pipe providing some external protection to the new plastic main.

In live insertion, the plastic pipe is inserted under gas pressure without the existing steel/cast iron main being first taken out of service. After the new plastic pipe is placed in position, tested, and connected, the old pipe is depressurized, vented, and rendered redundant.

In dead insertion, the pipe to be replaced is first taken out of service (thereby disconnecting customers) and new plastic pipe is inserted into the old pipe. The customers are re-connected to the new main. This work occurs under a no-gas condition.

A telephone survey of some North American gas utilities was done by the author in 1986, which showed that the majority of companies who had used insertion as a method of replacement adopted the dead-insertion method. The main advantage of live insertion is that customers are without gas for only a brief period, whereas for dead insertion, customers are without gas for many hours. Overall, the disadvantages and limitations of live insertion were found to be unacceptable to GUY. It was observed that many companies had looked into live insertion and decided to proceed with dead insertion. However, some companies preferred live insertion. At the time of the survey, two companies out of the ten surveyed were using live insertion. The advantages and disadvantages of live insertion are as follows:

Advantages	Disadvantages
1. Customer is without gas for only a short time while his service is being reconnected to the new main.	1. Crews are exposed to some gas leaking past the glands of the insertion fitting for a considerable time during the pipe-insertion operation. Explosive mixtures may form within the trench.
2. The need for a temporary gas supply is diminished.	2. Crews are exposed to the chemical isocyanate, which requires very careful handling and storage precautions. An occupational exposure limit (OEL) of 0.02 ppm is allowed under the occupational and safety rules. This chemical is a necessary component for foam making, which is used as a seal.

Advantages	Disadvantages
3. The insertion work can be terminated at any time without the need to provide temporary gas supply to the customers.	3. Chances of injection of dangerous chemicals into a live gas stream exist. This is possible if correct mixing of chemicals is not obtained during foam injection.
4. There will be fewer open trenches at any one time.	4. The foam that is used as a pressure seal is a combustible material. It releases great heat when on fire.
	5. With the equipment currently available, live insertion may only be carried out in low-pressure (172 kPa) and medium-pressure (210 kPa) systems and on pipe sizes up to 168 mm (NPS 6) diameter.
	6. In case of pipe failure during the testing of the inserted PE pipe, the possibility of nitrogen mixing with the natural gas and extinguishing customers' pilot lights exists. This drawback may be overcome by testing the pipe before insertion.
	7. Live insertion is more laborious and time consuming for insertion in steel pipe, as it involves first cutting a plug hole in the steel pipe, foaming, and then cutting out a window and then a cylindrical portion from the steel pipe in order to make sufficient room for the fusion equipment installation.
	8. If a sleeve pipe is to be provided in the steel pipe to protect the PE carrier pipe from scratches and freezing, then live insertion is much more difficult.

Having reviewed all facts, GUY selected the dead-insertion technique with partial direct burial where necessary, while at the same time further researching and developing the live-insertion technique.

3
Corrosion and Cathodic Protection

INTRODUCTION

The most common form of corrosion (see definitions on pages 38–43) attacking metallic structures is the electrochemical process by which the metal reacting with its environment converts into metallic oxides, thus forming more stable compounds. In doing so, it disintegrates and loses its structural strength. For the electrochemical cell to set up, two conditions are necessary:

1. There must be anodic and cathodic areas on the metal, i.e., there must be metal surfaces having different electrical potentials.
2. There must be a medium, such as an electrolyte, through which the current can flow so as to complete the circuit. An external connection between the metal surfaces is also necessary. This is provided by the pipe or the structure itself.

Since corrosion is dependent on the electron (current) flow, the higher the potential difference between the metal surfaces, the higher the corrosion rate. Similarly, if the circuit resistance is higher, the corrosion rate will be lower. This follows from the formula $I = V/R$ where I is current, V the voltage, and R the resistance. An electrochemical cell formed on a buried steel pipe is shown in Figure 3-1. Pipe-coating materials with high electrical resistance (dielectric strength), such as polyethylene and epoxies, are used to block the flow of hydrogen ions emitting from the anode (corroding surface) to the cathode (receiving surface). Corrosion cells set up where coating gets loose or damaged. Hydrogen ions emitting from the anodic areas travel through the medium (soil for underground steel structures) and reach the cathodic areas. At the same time, the electrons from the corroding surface travel via the pipe itself and meet at the receiving surface, thus completing the circuit. The build-up of hydrogen on the cathodic surface acts as a polarizer and tends to reduce the corrosion current flow. If the backfill material

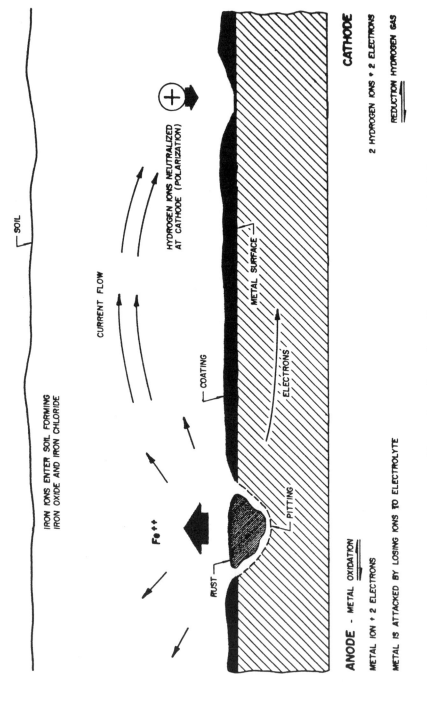

Figure 3-1. Corrosion cell on an unprotected steel pipe.

through which the hydrogen ions have to travel is of low resistivity, ion flow is faster and a greater risk of corrosion exists. Moist, acidic soils containing sodium chloride, ashes, and coke are inductive to an increased rate of corrosion. Conditions that promote corrosion cells setting up are many. They include:

1. **Dissimilar metals connected.** If two dissimilar metals are immersed in an electrolyte (soil for pipelines) and connected by a conductor (pipe itself), then the metal higher in the galvanic series (the one with higher electron energy) will go into solution, or corrode. There will be a flow of current from the corroding metal through the electrolyte to the non-corroding metal, which is lower in the galvanic series (Figure 3-2). For example, in the case of steel and copper connected in the ground, the steel becomes the anode, the copper becomes the cathode, and the loss of metal is from the steel.

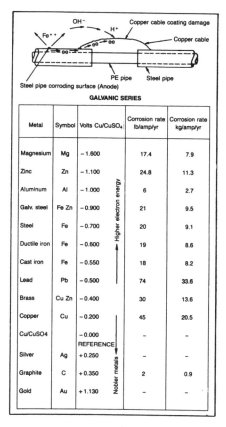

GALVANIC SERIES

Metal	Symbol	Volts Cu/CuSO₄	Corrosion rate lb/amp/yr	Corrosion rate kg/amp/yr
Magnesium	Mg	-1.600	17.4	7.9
Zinc	Zn	-1.100	24.8	11.3
Aluminum	Al	-1.000	6	2.7
Galv. steel	Fe Zn	-0.900	21	9.5
Steel	Fe	-0.700	20	9.1
Ductile iron	Fe	-0.600	19	8.6
Cast iron	Fe	-0.550	18	8.2
Lead	Pb	-0.500	74	33.6
Brass	Cu Zn	-0.400	30	13.6
Copper	Cu	-0.200	45	20.5
Cu/CuSO4		-0.000 REFERENCE	–	–
Silver	Ag	$+0.250$	–	–
Graphite	C	$+0.350$	2	0.9
Gold	Au	$+1.130$	–	–

Figure 3-2. Dissimilar metal corrosion and galvanic series.

2. **Dissimilar surface conditions.** Mill scale, scratches, gouges, welds, bends, localized stresses, imperfections of coating, or the age of pipe (i.e., new versus the old), may set up corrosion cells. Refer to Figure 3-3.

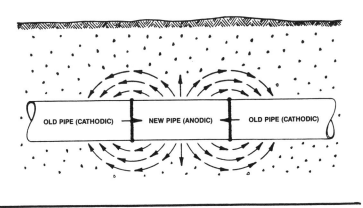

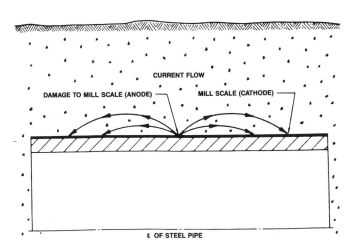

Figure 3-3. Galvanic cell due to dissimilar surface conditions.

3. **Dissimilar soil.** Metal surfaces lying within a low-resistivity soil will become anodic to that in a higher-resistivity soil. See Figure 3-4.
4. **Differential aeration of the electrolyte (soil).** This is also referred to as the oxygen-concentration cell. See Figure 3-5.
5. **Differential concentration of dissolved salts in the electrolyte.**

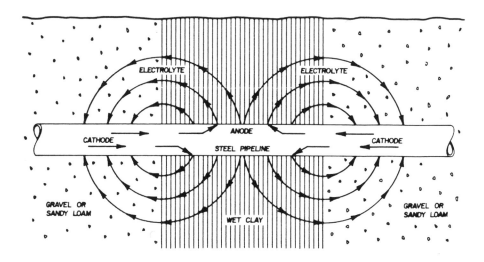

Figure 3-4. Galvanic cell due to dissimilar soils.

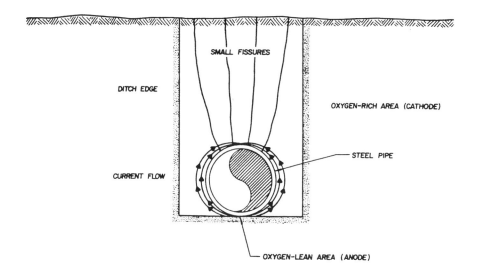

Figure 3-5. Oxygen concentration cell.

6. **Difference in moisture content of the soil.** 3, 4, 5, and 6 above fall under the definition of concentration-cell corrosion. The concentration cell consists of dissimilar electrolytes in contact with the metal (i.e., differences in pH, metal ion, or anion concentration, and in oxygen concentration).
7. **Microbiological.** Bacteria depolarizes the metal by removing the natural protective coating of the hydrogen.
8. **Stray current.** Foreign cathodic protection systems and other DC current systems (electric tracks, trolleys, street cars) may force the current to be picked up by a pipeline, which must be discharged at another location. Corrosion will occur where the current leaves the pipe. Refer to Figure 3-6.

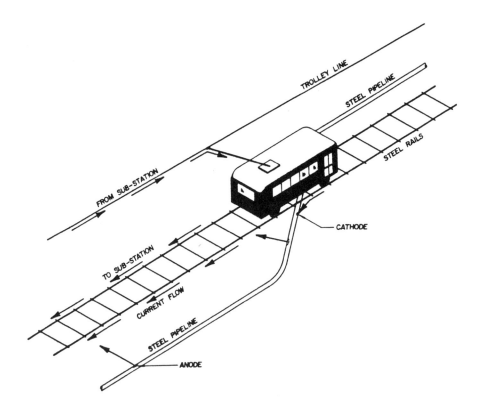

Figure 3-6. Stray-current corrosion.

CORROSION CONTROL

Corrosion can be minimized by coating the steel pipeline with a high-quality insulating material such as high-density polyethylene jacket or an epoxy film, and then providing it cathodic protection by means of sacrificial anodes or an impressed current system. However, there are a number of other considerations which may be applicable to a gas utility's pipeline system, which may be composed of a variety of materials such as steel, cast iron, copper, malleable iron fittings, bare and coated pipe, and underground and aboveground structures. The following considerations may be reviewed for new construction:

Layout

- While functional requirements and right-of-way routes determine the location of the pipeline, areas of higher corrosion susceptibility should be avoided if possible. This can be done by determining soil resistivity, moisture content, ground water level, rainfall, drainage, location of power lines, and soil pH values. In the context of gas distribution piping, such an evaluation is not done.

Selection of Materials

- Avoid using different metal types together. Where different metals must be used, isolate the metals electrically.
- Commercial steel and iron (cast and malleable) rate poorly, whereas ferrous alloys containing silicon, chromium, and nickel offer excellent corrosion resistance.

Joints

- Threaded joints, butt-welded joints with incomplete penetration, and lap-welded joints are subject to crevice corrosion and should be sealed.
- Avoid dissimilar metals.
- Coated pipe should be insulated from uncoated pipe.

Construction

- Avoid high localized stresses, i.e., the pipe should be laid on firm, sloping, or level bed. Wet gas lines should be sloped towards a collection point.
- Be sure there is adequate separation of crossing lines. The codes specify the minimum requirements. Insulate pipeline from all foreign structures.

- If excavated soil contains cinder, haul it away and backfill with clay or fillcrete. If this is not feasible, surround the pipe with 150 mm of sand. Remove rocks and other foreign material from the trench, as they may damage the coating or shield the pipe from protective current.

Water Treatment

- For the control of microorganisms around the underground piping, injection of a suitable algaecide such as chlorine will prevent microbiological corrosion.

CATHODIC PROTECTION

In Canada and the United States, cathodic protection is a code requirement for all gas pipelines. The purpose of cathodic protection is to mitigate corrosion by disrupting the flow of corrosion currents. This is accomplished by applying a DC current to the pipe via sacrificial anodes or from an impressed-current rectifier system. The first method consists of connecting and burying a material such as magnesium, which is less noble than the pipe metal, adjacent to the pipeline. In acting as an anode, the magnesium is consumed while the pipe is protected. When anodes are inadequate or impractical, an impressed current is used. Figures 3-7 and 3-8 show cathodic protection by sacrificial anodes and by impressed current (rectifier) methods.

The sacrificial (or galvanic) anode system purposefully establishes a dissimilar metal-corrosion cell strong enough to counteract corrosion cells normally existing on the pipeline. This is done by connecting and burying adjacent to the pipeline a strongly anodic metal, such as magnesium. This metal will corrode and in so doing will discharge current to the pipeline as shown in Figure 3-7.

The current available from a galvanic system is limited, so that when a large current is required or other factors dictate, a voltage from an outside power source may be "impressed" on the circuit between the pipeline and a replaceable anode bed.

It has been determined by corrosion engineers that steel and iron pipelines are satisfactorily protected when at least one of the following conditions are met:

1. The pipeline is at a negative potential on 0.85 volt with respect to the surrounding soil when the potential is measured with reference to a copper-copper sulfate ($Cu,Cu-SO_4$) electrode. Compensation for IR drops must be made in this measurement.

Example 3-1

Determine the theoretical time in years it will take for a corrosion leak to occur in a bare pipeline without cathodic protection. The pipeline is buried in homogeneous soil of soil resistivity 3,000 ohm·cm. The pipe wall thickness is 6.35 mm (0.25 in.).

Answer: For soil resistivity of 3,000 ohm·cm

$$CF = 3$$
$$WT = 6.35 \text{ mm}$$
$$Tp = \frac{12 \times 6.35}{3} = 25.4 \text{ years}$$

Note: No absolute formula has been developed to determine exactly when the first corrosion leak will appear as there are many variables involved. The above-mentioned formula provides an estimated value, and is based on observed cases.

USEFUL LIFE OF BURIED PIPELINE

A cathodically-unprotected steel pipeline requires increasing repairs after $1 \times Tp$ years have elapsed. Data on repair costs should be maintained for an economic and risk analysis. This analysis (see Chapter 1) is based on statistical information and provides an intelligent answer for scheduling the pipeline replacement. A rule of thumb is that when $2 \times Tp$ years have passed, the pipeline repair costs and leak frequency become too high, and when $3 \times Tp$ years have elapsed, the pipeline has failed completely. Because soil resistivity may change from spot to spot, time for first perforation and, therefore, the useful life may vary along the route of the pipeline. A pessimistic approach would be to calculate on the basis of the lowest soil resistivity. If the actual age of the pipeline at the time of first perforation (Tpa) is available, it should be used instead of Tp.

ANODE LIFE CALCULATION

The following formula may be used to estimate the useful life of a properly installed anode:

$$A_L = \frac{W \times E \times U}{CO \times CR}$$

where A_L = anode life in years
W = weight of anode in kg
E = anode efficiency
U = utilization factor = 0.85
CO = current output in amperes
CR = corrosion rate in kg/amp year

Anodes self-consume themselves to a degree; magnesium more so than zinc. The efficiency, E, of a magnesium anode is 0.5 while that of zinc is 0.95. The corrosion rate for zinc is 11.3 kg/amp/yr, while that of magnesium is 7.9 kg/amp/yr (see Figure 3-2). These numbers have already taken efficiency into account; hence, E should be dropped from the above formula if using numbers for CR from Figure 3-2.

Example 3-2

Calculate the useful life of a 2.3-kg zinc anode protecting an NPS 4 steel valve and two steel/PE transition fittings in a PE system in downtown Toronto. Refer to Figure 3-9.

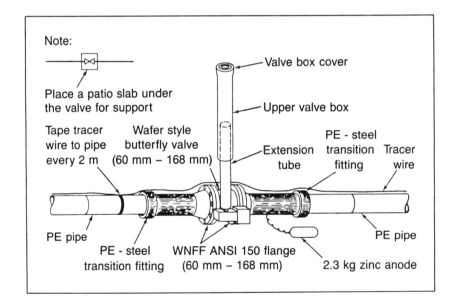

Figure 3-9. Typical buried steel valve in PE main.

It is safe to assume that the valve and the transition fittings will be well coated, requiring very little current. However, it is likely that the steel valve box (if used) will inadvertently contact the valve body and drain the anode. Because the metallic valve box is uncoated, its current requirement will be far greater than that of the coated transition fittings and the valve. Based on 0.3-volt drop across the coating between the steel pipeline and the soil, having a resistivity of 1,000 ohm·cm, the following current is required:

Bare pipe	11 mA/m²
Perfectly coated pipe having coating resistance of 10^{13} ohm/cm²	1.25×10^{-6} mA/m²
Observed coated pipe current requirement*	0.22 mA/m²

* About 2% of bare pipe requirement

Current requirements for the coated pipe are negligible (assuming no holidays), so in the example above, the current need for the bare steel may only be taken into consideration.

Estimated bare surface (measured) = 0.5 m²

Current requirement (current output CO) = 0.5 m² × 11 mA/m² = 5.5 mA

W = 2.3 kg zinc
U = 0.85
CR = 11.3 kg/A·yr from Figure 3-2

$$A_L = \frac{W \times U}{CO \times CR} = \frac{2.3 \times 0.85 \times 10^3}{5.5 \times 11.3}$$

Anode life = 31.46 yr

COATED STEEL PIPE IN BARE MAINS SYSTEM

It is a common practice to repair a section of leaking bare steel pipe by replacing it with a new, coated steel pipe. It is also customary to attach an anode to the new section of coated steel pipe. This practice of attaching an anode or a number of anodes may be fruitless unless the coated steel pipe is electrically insulated from the bare pipe. The anode will consume rapidly in trying to protect the bare pipes whose current requirements will be overwhelming. A rule of thumb is 1-kg magnesium (anode) is required to cathodically protect 1 m length of 168-mm dia. bare steel pipe in soil having 3,000 ohm·cm soil resistivity. This is only a rule of thumb; the actual requirements must be worked out based on surface area, soil resistivity, and other prevailing conditions. However, the installation of an anode or two on new coated section may give a degree of protection until the anodes get used up. If cathodic protection is not provided, galvanic cells may develop on the newly installed coated pipe and cause perforations in a time span much shorter than that experienced on the old bare pipe that it replaced. This will be true only if coating flaws are present. This emphasizes the need for careful handling of coated pipe and its holiday repairs prior to backfilling. In practice, 100% perfect coating is not achieved, hence the need for cathodic protection.

Table 3-2 gives the current and spacing requirement for 7.7-kg magnesium anode on bare steel pipeline. Bare-pipe current requirement is assumed 10.76 m A/m² (1 mA/ft²).

Table 3-2
Magnesium Anode Spacing On Bare Steel Pipe (m)

Soil Resistivity	Pipe Size (mm)						
	60	88	114	168	219	273	323
1,000	64	45	34	23	18	14	12
2,000	32	22.5	17	11.5	9	7	6
3,000	21	15	11.3	7.3	6	2.8	2.4
4,000	16	11.2	8.5	4.0	2.7		
5,000	12.8	9	6.8				
7,500	8.5	4	2.7				
10,000	6.4	2.7					
15,000	2.5						
20,000	2						
25,000	1.5						

Note: Anode efficiency drops significantly when anode spacing is less than 6 m; hence, more anodes are required below this spacing.

HOW MANY ANODES?

Anode needs are based on the type of anode, its size, its expected service life, its current output, the soil resistivity, the surface area of the protected structure, and the criteria of cathodic protection. Generally the criteria for cathodic protection are pre-established and are governed by regulatory codes and well-conceived corrosion practices.

Anode current output is dependent on the type and size of anodes and the soil resistivity. Current outputs of 2.3-kg zinc and 7.7-kg magnesium anodes connected to coated steel pipe and placed 1.5 m away, with a pipe to soil potential of − 0.85 V to Cu/Cu sulfate half cell are given in Table 3-3.

Table 3-3
Single Anode Current Output (mA)*

Soil Resistivity ohm·cm	Zinc	Magnesium
500	26	170
1,000	15	100
1,500	11	71
2,000	8	55
2,500	7	45
3,000	6	38
4,000	4	29
5,000	3	23
6,000	3	19
7,000	3	17
8,000	2	15

* If the anode is connected to bare steel pipe, the current output is higher.

Example 3-3

In the bare mains renewal program, all adjoining streets except a 9-block (9 × 180 m) length of 114-mm steel pipe were replaced by plastic pipe. This 9-block length was new yellow-jacketed steel pipe which was laid 2 years ago. It was decided to retest this pipe for higher maximum allowable operating pressure and, subject to satisfactory testing, to continue using this pipe. This pipe had one 7.7-kg magnesium anode attached to it at the time of initial installation. The soil resistivity is 5,000 ohm·cm and the anode is required to serve for a minimum of 15 years. Calculate if adequate protection is provided, and, if not, how many more anodes will be needed.

Data

Pipe surface area = $\pi \times 0.1143 \times 9 \times 180 = 582$ m^2

Current requirement = 0.22×10^{-3} A/m$^2 \times 582$ m$^2 = 128$ mA

Table 3-3 gives the current output of 7.7-kg magnesium anode as 23 mA; thus, previously, this line was not adequately protected.

$$\therefore \text{ No. of anodes required} = \frac{\text{Current requirement}}{\text{Current output/anode}}$$

$$= \frac{128}{23} = 5.5 \approx 6$$

Space five additional anodes, equidistant along the pipe route.

First anode—How much was left?

$$\text{Metal used up} = \text{current output} \times \text{yr} \times \frac{\text{weight}}{\text{A·yr}}$$

$$= 23 \times 10^{-3} \text{ A} \times 2 \text{ yr} \times \frac{7.9 \text{ kg}}{\text{A·yr}}$$

$$= 0.36 \text{ kg}$$

Anode left $= 7.7 - 0.36 = 7.34$ kg

Calculate anode life

Weight of anodes available taking utilization factor as 0.85

WU $= 5 \times 7.7 \times 0.85 + 0.85 \times 7.7 - 0.36 = 38.91$ kg

Current requirement = current output
$$= 127 \text{ mA}$$

Corrosion rate for magnesium CR $= 7.9$ kg/A·yr

$$\text{Anode life } A_L = \frac{W}{CO \times CR} = \frac{38.91 \text{ kg} \times \text{A} \times \text{yr}}{127 \times 10^{-3} \text{A} \times 7.9 \text{ kg}}$$

$$= 38.8 \text{ years is greater than 15 years, hence acceptable.}$$

Anode Spacing Calculations for Copper Tracer Wire for PE Pipe

A 2.3-kg (5-lb) zinc anode has the following:

Dimension = 10.2 cm dia. × 20.3 cm long, therefore r = 5.1 cm, L = 20.3 cm

Soil resistivity P = 3,000 ohm·cm assumed

Corrosion rate CR = 11.3 kg/A·yr

Potential difference of the driving voltage after polarization E = 1.1 − 0.85 (polarized) = 0.25 V

$$R = \text{anode resistance to soil} = \frac{0.159\ P}{L}\left(\log_e \frac{4L}{r} - 1\right)$$

$$R = \frac{0.159 \times 3,000}{20.3}\left(\log_e \frac{4 \times 20.3}{5.1} - 1\right)$$

−0.85V   Tracer wire
20.3 cm
10.2 cm dia.
−1.1V

= 63.51 ohms. Add 2 ohms for resistance at tracer wire, internal resistance, and cable resistance.

∴ R = 63.51 + 2 = 65.51 ohms

$$\text{Maximum current output } I = \frac{E}{R} = \frac{0.25}{65.51} \times 1,000 = 3.81 \text{ mA}$$

Take 2-km length of AWG 14 copper tracer wire. Current requirement based on 2% coating damage and 11 mA/m² current density = 2,000 × 2/100 × π × 1.628/1,000 × 11.

CO = 2.25 mA

Since this anode can supply 3.81 mA, the spacing of one 2.2-kg zinc anode every 2 km is fine.

$$\text{Anode life } A_L = \frac{W \times U}{CO \times CR} = \frac{2.3 \times 0.85 \times 10^3}{2.25 \times 11.3}\ \frac{\text{kg} \times A \times \text{yr}}{A \times \text{kg}} = 76.9 \text{ years}$$

IMPRESSED CURRENT/RECTIFIER

The impressed-current system is used when the current requirements are large. Generally, pipelines shorter than 3 km are not protected with rectifiers. Also, in areas of high electrical interference such as downtowns, where pavement is congested with other utilities and/or subway and trolley systems are present, the impressed-current system poses many problems. In the context of bare mains, which generally are located in high-interference areas, cathodic protection, where needed on coated steel pipe or on tracer wire, is therefore provided by anodes.

DEFINITIONS

Active: The negative direction of electrode potential. Also used to describe a metal that is corroding without significant influence of reaction product.

Aeration Cell (Oxygen Cell): *See* Differential Aeration Cell.

Anaerobic: Free of air or uncombined oxygen.

Anion: A negatively-charged ion that migrates through the electrolyte toward the anode under the influence of a potential gradient (e.g., Cl^- or OH^-).

Anode: The electrode of an electrolyte cell at which oxidation occurs. (Electrons flow away from the anode in the external circuit. It is usually at the electrode that corrosion occurs and metal ions enter solution.) A common anode reaction is $Zn \rightarrow Zn^{++} + 2$ electrons.

Anode Corrosion Efficiency: The ratio of the actual corrosion (weight loss) of an anode to the theoretical corrosion (weight loss) calculated by Faraday's Law from the quantity of electricity that has passed.

Anodic Polarization: The change of the electrode potential in the noble (positive) direction due to current flow. (*See* Polarization.)

Anodic Protection: Polarization to a more oxidizing potential to achieve a reduced corrosion rate by the promotion of passivity.

Anodizing: Oxide coating formed on a metal surface (generally aluminum) by an electrolytic process.

Anolyte: The electrolyte adjacent to the anode of an electrolytic cell.

Austenite: A face-centered cubic crystalline phase of iron-base alloys.

Auxiliary Electrode: An electrode commonly used in polarization studies to pass current to or from a test electrode. It is usually made from a noncorroding material.

Beach Marks: A term used to describe the characteristic fracture markings produced by fatigue crack propagation. Known also as clamshell, conchoidal, and arrest marks.

Bituminous Coating: Coal tar or asphalt-based coating.

Case Hardening: Hardening a ferrous alloy so that the outer portion, or case, is made substantially harder than the inner portion, or core. Typical processes are carburizing, cyaniding, carbon-nitrifying, nitriding, induction hardening, and flame hardening.

Cathode: The electrode of an electrolytic cell at which reduction is the principal reaction. (Electrons flow toward the cathode in the external circuit.) Typical cathodic processes are cations taking up electrons and being discharged, oxygen being reduced, and the reduction of an element or group of elements from a higher to a lower valence state.

Cathodic Corrosion: Corrosion resulting from a cathodic condition of a structure, usually caused by the reaction of an amphoteric metal with the alkaline products of electrolysis.

Cathodic Disbondment: The destruction of adhesion between a coating and its substrate by products of a cathodic reaction.

Cathodic Inhibitor: A chemical substance or mixture that prevents or reduces the rate of the cathodic reduction reaction.

Cathodic Polarization: The change of the electrode potential in the active (negative) direction caused by current flow. (*See* Polarization.)

Cathodic Protection: Reduction of corrosion rate by shifting the corrosion potential of the electrode toward a less oxidizing potential by applying an external electromotive force.

Catholyte: The electrolyte adjacent to the cathode of an electrolytic cell.

Cation: A positively charged ion which migrates through the electrolyte toward the cathode under the influence of a potential gradient.

Cell: Electrochemical system consisting of an anode and a cathode immersed in an electrolyte. The anode and cathode may be separate metals or dissimilar areas on the same metal. The cell includes the external circuit which permits the flow of electrons from the anode toward the cathode. (*See* Electrochemical Cell.)

Concentration Cell: An electrolytic cell, the electromotive force of which is caused by a difference in concentration of some component in the electrolyte. (This difference leads to the formation of discrete cathode and anode regions.)

Concentration Polarization: That portion of the polarization of a cell produced by concentration changes resulting from a passage of current through the electrolyte.

Contact Corrosion: A term (mostly used in Europe) to describe galvanic corrosion between dissimilar metal.

Continuity Bond: A metallic connection that provides electrical continuity between metal structures.

Corrosion: The deterioration of a material, usually a metal, by reaction with its environment.

Corrosion Fatigue: Fatigue-type cracking of metal caused by repeated or fluctuating stresses in a corrosive environment characterized by shorter life than would be encountered as a result of either the repeated or fluctuating stress alone or the corrosive environment alone.

Corrosion Potential (E_{corr}): The potential of a corroding surface in an electrolyte, relative to a reference electrode. Also called rest potential, open circuit potential, and freely corroding potential.

Corrosion Rate: The rate at which corrosion proceeds, expressed as either weight loss or penetration per unit time.

Corrosion Resistance: Ability of a metal to withstand corrosion in a given corrosion system.

Corrosivity: Tendency of an environment to cause corrosion in a given corrosion system.

Creep: Time-dependent strain occurring under stress.

Crevice Corrosion: A form of localized corrosion occurring at locations where easy access to the bulk environment is prevented, such as the mating surfaces of metals or assemblies of metal and nonmetal.

Critical Humidity: The relative humidity above which the atmospheric corrosion rate of some metals increases sharply.

Depolarization: The removal of factors resisting the flow of current in a cell.

Dielectric Shield: In a cathodic protection system, an electrically nonconductive material, such as a coating, plastic sheet, or pipe, that is placed between an anode and an adjacent cathode to avoid current wastage and to improve current distribution, usually on the cathode.

Differential Aeration Cell: An electrolytic cell, the electromotive force of which is due to a difference in air (oxygen) concentration at one electrode as compared with that at another electrode of the same material.

Electrochemical Cell: An electrochemical system consisting of an anode and a cathode in metallic contact and immersed in an electrolyte. (The anode and cathode may be different metals or dissimilar areas on the same metal surface.)

Electrochemical Equivalent: The weight of an element or group of elements oxidized or reduced at 100% efficiency by the passage of a unit quantity of electricity. Usually expressed as grams per coulomb.

Electrode: An electronic conductor used to establish electrical contact with an electrolytic part of a circuit.

Electrode Potential: The potential of an electrode in an electrolyte as measured against a reference electrode. (The electrode potential does not include any resistance losses in potential in either the solution or the external circuit. It represents the reversible work to move a unit charge from the electrode surface through the solution to the reference electrode.)

Electrolysis: The process that produces a chemical change in an electrolyte resulting from the passage of electricity.

Electrolyte: A chemical substance or mixture, usually liquid, containing ions that migrate in an electric field.

Electromotive Force Series (Emf Series): A list of elements arranged according to their standard electrode potentials, the sign being positive for elements whose potentials are cathodic to hydrogen and negative for those anodic to hydrogen.

Epoxy: Resin formed by the reaction of bisphenol and epichlorohydrin.

External Circuit: The wires, connectors, measuring devices, current sources, etc., that are used to bring about or measure the desired electrical conditions within the test cell. It is this portion of the cell through which electrons travel.

Ferrite: A body-centered cubic crystalline phase of iron-base alloys.

Fretting Corrosion: Deterioration at the interface of two contacting surfaces under load, accelerated by relative motion between them of sufficient amplitude to produce slip.

Galvanic Anode: A metal that, because of its relative position in the galvanic series, provides sacrificial protection to metals that are more noble in the series when coupled in an electrolyte.

Galvanic Corrosion: Corrosion associated with the current resulting from the electrical coupling of dissimilar electrodes in an electrolyte.

Galvanic Current: The electric current that flows between metals or conductive nonmetals in a galvanic couple.

Galvanic Series: A list of metals arranged according to their corrosion potentials in a given environment.

General Corrosion: A form of deterioration that is distributed more or less uniformly over a surface.

Graphitic Corrosion: Deterioration of gray cast iron in which the metallic constituents are selectively leached or converted to corrosion products leaving the graphite intact.

Ground Bed: A buried item, such as junk steel or graphite rods, that serves as the anode for the cathodic protection of pipelines or other buried structures.

Half-Cell: A pure metal in contact with a solution of known concentration of its own ion, at a specific temperature develops a potential which is characteristic and reproducible; when coupled with another half-cell, an overall potential develops which is the sum of both half-cells.

Holiday: Any discontinuity or bare spot in a coated surface.

Hydrogen Overvoltage: Overvoltage associated with the liberation of hydrogen gas.

Impressed Current: Direct current supplied by a device employing a power source external to the electrode system of a cathodic protection installation.

Inhibitor: A chemical substance or combination of substances that, when present in the environment, prevents or reduces corrosion without significant reaction with the components of the environment.

Ion: An electrically charged atom or group of atoms.

Mill Scale: The heavy oxide layer formed during hot fabrication or heat treatment of metals.

Noble Metal: A metal that occurs commonly in nature in the free state. Also a metal or alloy whose corrosion products are formed with a low negative or positive free-energy charge.

Open-Circuit Potential: The potential of an electrode measured with respect to a reference electrode or another electrode when no current flows to or from it. (*See* Corrosion Potential.)

Overvoltage: The change in potential of an electrode from its equilibrium or steady-state value when current is applied.

Oxidation: Loss of electrons by a constituent of a chemical reaction. (Also refers to the corrosion of a metal that is exposed to an oxidizing gas at elevated temperatures.)

Oxygen Concentration Cell: *See* Differential Aeration Cell.

Passive-Active Cell: A cell, the electromotive force of which is caused by the potential difference between a metal in an active state and the same metal in a passive state.

Patina: The coating, usually green, which forms on the surface of metals such as copper and copper alloys exposed to the atmosphere. Also used to describe the appearance of a weathered surface of any metal.

pH: A measure of hydrogen ion activity defined by

$$pH = Log_{10} \frac{1}{aH^+}$$

where aH^+ = hydrogen ion activity = molar concentration of hydrogen ions multiplied by the mean ion activity coefficient. A pH value of seven is neutral, low numbers are acidic, higher numbers are alkaline.

Pits, Pitting: Localized corrosion of a metal surface that is confined to a small area and takes the form of cavities.

Pitting Factor: The ratio of depth of the deepest pit resulting from corrosion divided by the average penetration as calculated from weight loss.

Polarization: The deviation from the open circuit potential of an electrode resulting from the passage of current.

Reduction: Gain of electrons by a constituent of a chemical reaction, as when copper is electroplated on steel from a copper sulfate solution (opposite of oxidation).

Rust: Corrosion product consisting primarily of hydrated ion oxide; a term properly applied only to iron and ferrous alloys.

Stray Current Corrosion: Corrosion resulting from direct current flow through paths other than the intended circuit. For example, by an extraneous current in the earth.

REFERENCES

1. *Corrosion Basics*, National Association of Corrosion Engineers (NACE), Houston, Texas, 1984.
2. *NACE Glossary of Corrosion-Related Terms*, National Association of Corrosion Engineers, Houston, Texas, 1986.

4
Equivalent Pipe Sizes

Determining future operating pressure in a major replacement project is a crucial question, as it forms the basis for sizing the system and affects the cost. This chapter considers the options available for choosing an operating pressure for the replacement mains. It discusses the impact of size combinations on capacity, operational considerations, and code requirements to help you arrive at a decision for the new operating pressure. (Also review Chapter 8 for more information on this topic.)

SCOPE

This chapter uses the Institute of Gas Technology, Chicago (IGT) gas-flow formula to determine equivalent pipe sizes for replacing existing low-pressure—1.72 kPa (¼ psi)—and medium-pressure—105 kPa (15 psi)—steel pipe with polyethylene (PE) pipe for operation at 105 kPa (15 psi), 210 kPa (30 psi), 415 kPa (60 psi), and 550 kPa (80 psi) pressure.

COST REDUCTION

The intent of this chapter is to explore ways to minimize costs while still providing a safe and reliable distribution system. It is well known that:

1. Lower pressure operations require larger sized mains, which are more expensive to construct.
2. Polyethylene mains up to NPS (nominal pipe size) 6 are cheaper to construct than the same size steel mains.

3. The insertion method is cheaper than the direct burial method. Polyethylene pipe lends itself easily to the insertion method.

By adopting the insertion technique, costs can be reduced. Costs can be further lowered by increasing the pressure, and therefore using smaller diameter pipes.

However, any increase in the system pressure may affect the company's operating practices, such as those on tooling, material, equipment, and emergency procedures. Therefore, it is important to thoroughly consider the implications of a pressure increase. This chapter considers these matters and recommends appropriate actions.

ASSUMPTIONS AND METHODOLOGY

It is assumed that the current practice of using PE (polyethylene) materials for gas distribution piping is acceptable. The current trend is to operate distribution systems up to the maximum allowed pressure under the codes. Pressures of up to 550 kPa (80 psi) are considered. This is the present limit of pressure for SDR 11 PE pipe having hydrostatic design basis (HDB) stress of 8.62 MPa (1,350 psi).

In this chapter, relative carrying capacities of various sizes of plastic pipe that could be used to replace existing steel lines are calculated. This is done for various operating pressures. Calculations are based on IGT's flow equation. For flow comparison, a pressure drop of 10% is assumed for both the existing steel main and for the replacement plastic mains. The atmospheric pressure is taken as 101.325 kPa (14.7 psi) for this study. Although actual atmospheric pressure differs slightly from site to site, its effect on this analysis is insignificant. In selecting an operating pressure, the following criteria were applied:

1. The capacity of the replacement pipe should be equal or greater than that of the existing mains.
2. Cross-sectional areas of the replacement pipe are at least 40% of the cross-sectional area of flow of the existing main.

IGT GAS EQUATION

The IGT gas equation simplifies to:

$$P_1^2 - P_2^2 = R \, Q^{1.8} \, L$$

where $R = 0.3735 \times \dfrac{(9.9856 \times 10^{-2})}{D^{4.8}}$ $(kPa)^2$

P_1 = initial pressure (kPa)
P_2 = final pressure (kPa)
L = length of pipe (m)
D = inside diameter of pipe (in.)

Calculated values of R are:

Type	Size (NPS)	D (in.)	R
ST	3/4	0.824	9.445 E − 02
	1¼	1.380	7.948 E − 03
	2	2.157	9.315 E − 04
	3	3.250	1.302 E − 04
	4	4.250	3.593 E − 05
	6	6.313	5.377 E − 06
	8	8.249	1.489 E − 06
	10	10.374	4.956 E − 07
	12	12.312	2.178 E − 07
PE	½ CTS	0.510	9.446 E − 01
	3/4	0.861	9.650 E − 02
	1¼	1.328	9.557 E − 03
	1½	1.554	4.495 E − 03
	2	1.943	1.538 E − 03
	2½	2.352	6.149 E − 04
	3	2.864	2.389 E − 04
	4	3.682	7.153 E − 05
	6	5.425	1.113 E − 05
	8	7.055	3.154 E − 06
	10	8.75	1.122 E − 06

The values of R given above are based on gas gravity of 0.6, base temperature of 10°C (50°F), and base pressure of 95.22 kPa. These values have not been adjusted as their effect is insignificant on the final results.

Let suffix 2 denote conditions at higher pressure and let suffix 1 denote conditions at lower pressure (1.72 kPag).

From IGT equation:

$$P_1^2 = R_1 \, Q_1^{1.8} \, L_1 \tag{4-1}$$

$$P_2^2 = R_2 \, Q_2^{1.8} \, L_2 \tag{4-2}$$

Because we are considering the same length of pipe, L_1 and L_2 will cancel out.

$\therefore$ dividing Equation 4-2 by Equation 4-1 gives

$$\left(\frac{Q_2}{Q_1}\right)^{1.8} = \frac{\Delta P_2^2}{\Delta P_1^2} \times \frac{R_1}{R_2}$$

or

$$\frac{Q_2}{Q_1} = \left(\frac{\Delta P_2^2}{\Delta P_1^2}\right)^{0.556} \times \left(\frac{R_1}{R_2}\right)^{0.556} \tag{4-3}$$

If we set up tables of values of:

$$\left(\frac{\Delta P_2^2}{\Delta P_1^2}\right)^{0.556} \text{ and of } \left(\frac{R_1}{R_2}\right)^{0.556}$$

for various sizes of pipe and for various pressure classes, we can compare the flow capacities of respective alternatives. So, let us first develop the required tables (Tables 4-1 and 4-2).

Table 4-1
Nominal Dimensions of Steel and PE Pipe

Size NPS	Steel Pipe ID* (in.)	SDR 11, PE	
		Pipe ID (in.)	Pipe OD (in.)
12	12.313		
10	10.374	8.75	10.75
8	8.249	7.055	8.625
6	6.313	5.425	6.625
4	4.25	3.68	4.5
3	3.25	2.86	3.5
2	2.157	1.955	2.375
1 1/4	1.380	1.36	1.66
3/4	0.828	0.870	1.05
1/2 CTS	—	0.510	0.625

* Based on commonly-used wall thicknesses for distribution pipe.
ID = inner diameter
OD = outer diameter

Table 4-2
R Ratios

Steel Size/PE Insert Size (NPS)	R_1/R_2	$(R_1/R_2)^{0.556}$
12/8	$2.178 \times 10^{-7}/3.154 \times 10^{-6} = 0.069$	0.226
12/6	$2.176 \times 10^{-7}/1.113 \times 10^{-5} = 0.0195$	0.112
12/4	$2.178 \times 10^{-7}/7.153 \times 10^{-5} = 0.00304$	0.0398
12/3	$2.178 \times 10^{-7}/2.389 \times 10^{-4} = 0.000911$	0.0204
10/6	$4.956 \times 10^{-7}/1.113 \times 10^{-5} = 0.0445$	0.177
10/4	$4.956 \times 10^{-7}/7.153 \times 10^{-5} = 0.00692$	0.063
10/3	$4.956 \times 10^{-7}/2.389 \times 10^{-4} = 0.00207$	0.0322
10/2	$4.956 \times 10^{-7}/1.538 \times 10^{-3} = 0.0003222$	0.0114
8/6	$1.489 \times 10^{-6}/1.113 \times 10^{-5} = 0.1337$	0.326
8/4	$1.489 \times 10^{-6}/7.153 \times 10^{-5} = 0.0208$	0.116
8/3	$1.489 \times 10^{-6}/2.389 \times 10^{-4} = 0.00623$	0.059
8/2	$1.489 \times 10^{-6}/1.538 \times 10^{-3} = 0.00968$	0.021
6/4	$5.377 \times 10^{-6}/7.153 \times 10^{-5} = 0.075$	0.237
6/3	$5.377 \times 10^{-6}/2.389 \times 10^{-4} = 0.0225$	0.121
6/2	$5.377 \times 10^{-6}/1.538 \times 10^{-3} = 0.00349$	0.043
6/1 1/4	$5.377 \times 10^{-6}/9.557 \times 10^{-3} = 0.0005626$	0.01559
4/3	$3.593 \times 10^{-5}/2.389 \times 10^{-4} = 0.150$	0.348
4/2	$3.593 \times 10^{-5}/1.538 \times 10^{-3} = 0.0233$	0.123
4/1 1/4	$3.593 \times 10^{-5}/9.557 \times 10^{-3} = 0.00375$	0.0448
3/2	$1.302 \times 10^{-4}/1.538 \times 10^{-3} = 0.0846$	0.253
3/1 1/4	$1.302 \times 10^{-4}/9.557 \times 10^{-3} = 0.01362$	0.0917
2/1 1/4	$9.315 \times 10^{-4}/9.557 \times 10^{-3} = 0.09746$	0.274
2/3/4	$9.315 \times 10^{-4}/7.650 \times 10^{-2} = 0.0119$	0.085
2/1 1/2 CTS	$9.315 \times 10^{-4}/9.446 \times 10^{-1} = 0.000986$	0.0213
1 1/4/3/4	$7.948 \times 10^{-3}/7.65 \times 10^{-2} = 0.1038$	0.283
1 1/4/1 1/2 CTS	$7.948 \times 10^{-3}/9.446 \times 10^{-1} = 0.00841$	0.07019
3/4/1 1/2 CTS	$9.445 \times 10^{-2}/9.446 \times 10^{-1} = 0.0999$	0.2779

ΔP^2 Simplification

For relative comparison of flow capacity, assume a 10% pressure drop in the gauge pressure in a given length for both the old steel pipe and the new pipe to be installed.

Let P_1 = absolute pressure at the start of the given length of pipe
   = $Pg_1 + Pa$

where Pg_1 = gauge pressure at start
   Pa = atmospheric pressure

$$\therefore P_1^2 = Pg_1^2 + Pa^2 + 2 Pg_1 Pa$$

assume P_2 = absolute pressure at the end
$$= 0.9 Pg_1 + Pa$$

$$\therefore P_2^2 = 0.81 Pg_1^2 + Pa^2 + 1.8 Pg_1 Pa$$

$$\therefore \Delta P^2 = P_1^2 - P_2^2 = 0.19 Pg_1^2 + 0.20 Pg_1 Pa$$

Suffix 1 may be dropped to make it a general formula.

$$\therefore \Delta P^2 = 0.19 Pg^2 + 0.20 Pg \times Pa \qquad \text{Hayat's Equation (1)}$$

ΔP^2 for LP—1.72 kPag ($^1/_4$ psig)

$$\Delta P^2 = 0.19 Pg^2 + 0.2 Pg Pa$$
$$= 0.19 \times 1.72^2 + 0.2 \times 1.72 \times 101.325$$
$$= 0.562 + 34.85 = 35.417$$

ΔP^2 for MP—105 kPag (15 psig)

$$\Delta P^2 = 0.19 \times 105^2 + 0.2 \times 105 \times 101.325$$
$$= 2{,}094.75 + 2{,}127.825 = 4{,}222$$

ΔP^2 for IP—210 kPag (30 psig)

$$\Delta P^2 = 0.19 \times 210^2 + 0.2 \times 210 \times 101.325$$
$$= 12{,}634$$

ΔP^2 for IP—415 kPag (60 psig)

$$\Delta P^2 = 0.19 \times 415^2 + 0.2 \times 415 \times 101.325$$
$$= 32{,}722 + 8{,}409.9 = 41{,}132.7$$

$$\Delta P^2 \text{ for IP—550 kPag (80 psig)}$$

$$\Delta P^2 = 0.19 \times 550^2 + 0.2 \times 550 \times 101.325$$
$$= 57{,}475 + 11{,}145.75 = 68{,}620.75$$

Table 4-3
LP Capacity Comparison If Operating Pressure 105 kPa (15 psi)

$$\frac{Q_{105}}{Q_{1.72}} = \left(\frac{\Delta P^2_{105}}{\Delta P^2_{1.72}}\right)^{0.556} \times \left(\frac{R_{Steel}}{R_{PE}}\right)^{0.556} = 14.27 \text{ R ratio}$$

Select values from R ratio and ΔP^2, see pages 48–49.

Steel Size/PE Insert Size (NPS)	Capacity Ratio	Q_2/Q_1
12/8	14.27 × 0.226	= 3.22
12/6	14.27 × 0.112	= 1.6
12/4	14.27 × 0.0398	= 0.56
12/3	14.27 × 0.0204	= 0.291
10/6	14.27 × 0.177	= 2.52
10/4	14.27 × 0.063	= 0.899
10/3	14.27 × 0.0322	= 0.459
10/2	14.27 × 0.0114	= 0.162
8/6	14.27 × 0.326	= 4.65
8/4	14.27 × 0.116	= 1.65
8/3	14.27 × 0.059	= 0.84
8/2	14.27 × 0.021	= 0.299
6/4	14.27 × 0.237	= 3.38
6/3	14.27 × 0.121	= 1.726
6/2	14.27 × 0.043	= 0.613
6/1 1/4	14.27 × 0.01559	= 0.222
4/3	14.27 × 0.3487	= 4.975
4/2	14.27 × 0.123	= 1.755
4/1 1/4	14.27 × 0.0448	= 0.639
3/2	14.27 × 0.253	= 3.61
3/1 1/4	14.27 × 0.0917	= 1.30
2/1 1/4	14.27 × 0.274	= 3.9
2/3/4	14.27 × 0.085	= 1.21
2/1/2 CTS	14.27 × 0.0213	= 0.303
1 1/4/3/4	14.27 × 0.283	= 4.038
1 1/4/1/2 CTS	14.27 × 0.0719	= 1.02
3/4/1/2 CTS	14.27 × 0.277	= 3.95

Table 4-4
LP Capacity Comparison If Operating Pressure 210 kPa (30 psi)

$$\frac{Q_{210}}{Q_{1.72}} = \left(\frac{\Delta P^2_{210}}{\Delta P^2_{1.72}}\right)^{0.556} \times \left(\frac{R_{Steel}}{R_{PE}}\right)^{0.556} = 26.248 \text{ R ratio}$$

Steel Size/PE Insert Size (NPS)	Capacity Ratio	Q_2/Q_1
12/8	26.248 × 0.226	= 5.93
12/6	26.248 × 0.112	= 2.93
12/4	26.248 × 0.0398	= 1.04
12/3	26.248 × 0.0204	= 0.535
10/6	26.248 × 0.177	= 4.64
10/4	26.248 × 0.063	= 1.65
10/3	26.248 × 0.0322	= 0.845
10/2	26.248 × 0.0114	= 0.299
8/6	26.248 × 0.326	= 8.55
8/4	26.248 × 0.116	= 3.04
8/3	26.248 × 0.059	= 1.54
8/2	26.248 × 0.021	= 0.551
6/4	26.248 × 0.237	= 6.22
6/3	26.248 × 0.121	= 3.17
6/2	26.248 × 0.043	= 1.13
6/1¼	26.248 × 0.01559	= 0.409
4/3	26.248 × 0.3487	= 9.15
4/2	26.248 × 0.123	= 3.228
4/1¼	26.248 × 0.0448	= 1.175
3/2	26.248 × 0.253	= 6.64
3/1¼	26.248 × 0.0917	= 2.4
2/1¼	26.248 × 0.274	= 7.19
2/¾	26.248 × 0.085	= 2.23
2/1½ CTS	26.248 × 0.0213	= 0.56
1¼/¾	26.248 × 0.283	= 7.43
1¼/1½ CTS	26.248 × 0.0719	= 1.88
¾/1½ CTS	26.248 × 0.277	= 7.27

Table 4-5
LP Capacity Comparison If Operating Pressure 415 kPa (60 psi)

$$\frac{Q_{410}}{Q_{1.72}} = \left(\frac{\Delta P^2_{415}}{\Delta P^2_{1.72}}\right)^{0.556} \times \left(\frac{R_{Steel}}{R_{PE}}\right)^{0.556} = \left(\frac{41,132.7}{35.417}\right)^{0.556} \times \text{R ratio} = 53.148 \text{ R ratio}$$

Steel Size/PE Insert Size (NPS)	Capacity Ratio	Q_2/Q_1
12/8	53.148 × 0.226	= 12.01
12/6	53.148 × 0.112	= 5.96

(Table 4-5 continued on next page)

Table 4-5 Continued
LP Capacity Comparison If Operating Pressure 415 kPa (60 psi)

Steel Size/PE Insert Size (NPS)	Capacity Ratio	Q_2/Q_1
12/4	53.148 × 0.0398	= 2.12
12/3	53.148 × 0.0204	= 1.084
10/6	53.148 × 0.177	= 9.42
10/4	53.148 × 0.063	= 3.3
10/3	53.148 × 0.0322	= 1.7
10/2	53.148 × 0.0114	= 0.605
8/6	53.148 × 0.326	= 17.3
8/4	53.148 × 0.116	= 6.17
8/3	53.148 × 0.059	= 3.157
8/2	53.148 × 0.021	= 1.116
6/4	53.148 × 0.237	= 12.6
6/3	53.148 × 0.121	= 6.447
6/2	53.148 × 0.043	= 2.28
6/1 1/4	53.148 × 0.01159	= 0.616
4/3	53.148 × 0.3487	= 18.53
4/2	53.148 × 0.123	= 6.58
4/1 1/4	53.148 × 0.0448	= 2.38
3/2	53.148 × 0.253	= 13.46
3/1 1/4	53.148 × 0.0917	= 4.87
2/1 1/4	53.148 × 0.274	= 14.56
2/3/4	53.148 × 0.085	= 4.53
2/1/2 CTS	53.148 × 0.0213	= 1.13
1 1/4/3/4	53.148 × 0.283	= 15.09
1 1/4/1/2 CTS	53.148 × 0.07019	= 3.73
3/4/1/2 CTS	53.148 × 0.277	= 14.77

Table 4-6
LP Capacity Comparison If Operating Pressure 550 kPa (80 psi)

$$\frac{Q_{550}}{Q_{1.72}} = \left(\frac{\Delta P^2_{550}}{\Delta P^2_{1.72}}\right)^{0.556} \times \left(\frac{R_{Steel}}{R_{PE}}\right)^{0.556} = \left(\frac{68{,}620.75}{35.417}\right)^{0.556} \times \text{R ratio} = 67.25 \times \text{R ratio}$$

Steel Size/PE Insert Size (NPS)	Capacity Ratio	Q_2/Q_1
12/8	67.25 × 0.226	= 15.2
12/6	67.25 × 0.112	= 7.53
12/4	67.25 × 0.0398	= 2.67
12/3	67.25 × 0.0204	= 1.37
10/6	67.25 × 0.177	= 11.90
10/4	67.25 × 0.063	= 4.24
10/3	67.25 × 0.0322	= 2.16
10/2	67.25 × 0.0114	= 0.766

Table 4-6 Continued
LP Capacity Comparison If Operating Pressure 550 kPa (80 psi)

Steel Size/PE Insert Size (NPS)	Capacity Ratio Q_2/Q_1	
8/6	67.25×0.326	$= 21.92$
8/4	67.25×0.116	$= 7.8$
8/3	67.25×0.059	$= 3.96$
8/2	67.25×0.021	$= 1.41$
6/4	67.25×0.237	$= 15.93$
6/3	67.25×0.121	$= 8.14$
6/2	67.25×0.043	$= 2.89$
6/1¼	67.25×0.01559	$= 1.05$
4/3	67.25×0.3487	$= 23.45$
4/2	67.25×0.123	$= 8.27$
4/1¼	67.25×0.0448	$= 3.01$
3/2	67.25×0.253	$= 17.01$
3/1¼	67.25×0.0917	$= 6.16$
2/1¼	67.25×0.274	$= 18.42$
2/¾	67.25×0.085	$= 5.71$
2/½ CTS	67.25×0.0213	$= 1.43$
1¼/¾	67.25×0.283	$= 19.03$
1¼/½ CTS	67.25×0.07019	$= 4.72$
¾/½ CTS	67.25×0.2779	$= 18.69$

Table 4-7
LP Capacity Comparison—Q_2/Q_1 Summary Sheet

Steel Size	Operating Pressure (kPa)				
PE size	1.72 (1/4 psi)	105* (15 psi)	210 (30 psi)	410 (60 psi)	550 (80 psi)
12/8	0.226	3.22	5.93	12.01	15.2
*12/6	0.112	1.6	2.93	5.96	7.53
12/4	0.0398	0.56	1.04	2.11	2.67
12/3	0.0204	0.291	0.535	1.084	1.37
*10/6	0.117	2.52	4.64	9.42	11.9
10/4	0.063	0.899	1.65	3.3	4.24
10/3	0.0322	0.459	0.845	1.7	2.16
10/2	0.0114	0.162	0.299	0.605	0.766
8/6	0.326	4.65	8.55	17.3	21.92
*8/4	0.116	1.65	3.04	6.17	7.8
8/3	0.059	0.84	1.54	3.157	3.96
8/2	0.021	0.299	0.551	1.116	1.41
6/4	0.237	3.38	6.22	12.6	15.93
*6/3	0.121	1.726	3.17	6.447	8.14

* Recommended sizes and pressure combinations, assuming freedom from freezing, and adequate capacities.

(Table 4-7 continued on next page)

Table 4-7 Continued
LP Capacity Comparison—Q_2/Q_1 Summary Sheet

Steel Size	Operating Pressure (kPa)				
PE size	1.72 (1/4 psi)	105* (15 psi)	210 (30 psi)	410 (60 psi)	550 (80 psi)
6/2	0.043	0.613	1.13	2.28	2.89
6/1 1/4	0.01159	0.222	0.409	0.616	1.05
4/3	0.348	4.975	9.15	18.53	23.45
*4/2	0.123	1.755	3.228	6.58	8.27
4/1 1/4	0.0448	0.639	1.175	2.38	3.01
3/2	0.253	3.61	6.64	13.46	17.01
*3/1 1/4	0.917	1.30	2.4	4.87	6.16
*2/1 1/4	0.274	3.9	7.19	14.56	18.42
Services					
2/3/4	0.85	1.21	2.23	4.53	5.71
2/1/2 CTS	0.213	0.303	0.56	1.13	1.43
*1 1/4/3/4	0.283	4.038	7.43	15.09	19.03
1 1/4/1/2 CTS	0.07	1.02	1.88	3.73	4.72
*3/4/1/2 CTS	0.278	3.95	7.27	14.77	18.69

* Recommended sizes and pressure combinations, assuming freedom from freezing, and adequate capacities.

Table 4-8
Summary of Pipe Size Selection for Replacement of LP 1.72 kPa (1/4 psi) System

PE pipe sizes listed will give equal or higher capacities than that of existing steel-pipe sizes.

Existing Steel Pipe Size	Pressure 105 kPa (15 psi)	210 kPa (30 psi)	410 kPa (60 psi)	550 kPa (80 psi)
12	6	4	3	3
10	6	4	3	3
8	4	3	2	2
6	3	2	2	1 1/4
4	2	1 1/4	1 1/4	1 1/4
3	1 1/4	1 1/4	1 1/4	1 1/4
2	1 1/4	1 1/4	1 1/4	1 1/4
	(2 sizes down)	(3 sizes down)	(4 sizes down)	(4 sizes down)

It is interesting to note that sizes for 105 kPa (15 psi) are 2 sizes smaller than the casing pipe; for 210 kPa (30 psi), 3 sizes down; for 440 kPa (60 psi), 4 sizes down; for 550 kPa (80 psi), 4 sizes down. The only exceptions to the rule are 12/6 for 105 kPa (15 psi), and 6/2 for 410 kPa (60 psi) pressure.

* Note: Minimum PE main size is chosen to be NPS 1 1/4.

Check by Weymouth Formula

Let us stop and check some of the values calculated by the Weymouth formula using the same basic parameters.

Weymouth simplified: $Q = c \dfrac{d^{8/3}}{\sqrt{L}} \dfrac{p_1^2 - p_2^2}{}$

$$\therefore \frac{Q_2}{Q_1} = \left(\frac{d_2}{d_1}\right)^{2.667} \times \left(\frac{\Delta p_2^2}{\Delta p_1^2}\right)^{0.5}$$

Comparison of flow capacity between NPS 4 steel LP main and NPS 2 PE at 550 kPa (80 psi), with 10% pressure drop in each case.

$$\frac{Q_2}{Q_1} = \left(\frac{1.955}{4.25}\right)^{2.667} \times \left(\frac{68,620.75}{35.417}\right)^{0.5}$$

$$= (0.46)^{2.667} \times (1,937.50)^{0.5}$$

$$= 0.126 \times 44 = 5.54$$

Use of IGT formula gives this ratio as 8.27. See Table 4-6.
Let us compare these numbers for 105 kPa (15 psi).

$$\frac{Q_2}{Q_1} = \left(\frac{1.955}{4.25}\right)^{2.667} \times \left(\frac{4,222}{35.417}\right)^{0.5}$$

$$= 0.126 \times 10.9 = 1.375$$

IGT method gives this value = 1.755. The magnitude of difference is smaller at lower pressures.

Since the purpose of this chapter is not to explore the virtues of various gas flow formulas, the impact of using the IGT formula versus the Weymouth formula will be ignored. The purpose of the above analysis is to determine the effect on cost by choosing one or the other operating pressure. It should be noted that the Weymouth formula in the simplified version presented is less accurate, as it ignores the super-compressibility factor. That is the reason for a wider gap in answers at higher pressures. We will use the IGT equation.

CONSIDERATIONS FOR SIZE SELECTION

Cost. Since the construction cost is directly related to the size of the main, the smaller the size, the lower the cost.

Availability. CSA-approved pipe, fittings, and equipment for polyethylene pipe are readily available in Canada in sizes ranging from 1/2 in. CTS (copper tube size) to NPS 6. Even larger sized pipe, fittings, and equipment are available in the United States and Europe.

Handling. The smaller the pipe, the easier it is to handle. Those with slip lining experience recommend that the size of the insert pipe should be kept two sizes smaller than the casing (steel) pipe. This ensures that the insert pipe will, during insertion, move through the casing easily and without damage. This recommendation sets the higher limit of the size of the insert.

Freezing. The ground water will have easy ingress to the casing pipe, and the likelihood that under some circumstances water may completely fill a section of the main and, in cold climates, freeze to form an ice cylinder around the polyethylene pipe exists. Theoretically, it is possible for the ice formation to exert pressure on the plastic pipe to such an extent that the flow of gas is partially or completely blocked. Water expands by approximately 10% upon freezing.

Example 4-1

NPS 12 steel casing, cross-sectional area $= \frac{\pi}{4} \times 12.313^2 = 119$ in.^{2}

NPS 3 PE insert, cross-sectional area $= \frac{\pi}{4} \times 2.86^2 = 6.42$ in.^{2}

X^n area available to water $= 119 - 6.42 = 112.57$ in.^{2}

If the main is full of water, the ice will require a cross-sectional area of $112.57 \times 1.1 = 123$ in.^{2} upon freezing, i.e., an additional 10.43 in.^{2}, thus flattening the plastic main completely as well as exerting radial stresses in the steel casing.

It is therefore prudent to keep the size of the insert well above the 30% ratio. For mains, set this ratio near 50%. This will result in a loss of capacity, upon water freezing in the casing, of no more than 13%. For services, this limit may be set lower, to about 30%, incurring a maximum possible capacity loss of approximately 30%. This calculation is based on the assumption that the casing water temperatures remain uniform over the entire length, and thus the freezing action is uniform. Note: Try the following sizing limitations:

Lower limit of insert: Main—50% of casing x^n area
Service—30% of casing x^n area

Tables 4-9 and 4-10 give 30% and 50% areas of casing pipe.

Table 4-9
Pipe Cross-Sectional Area

	Steel Pipe		PE Pipe	
Size NPS	ID (in.)	Internal x^n area (in.^{2})	OD (in.)	Total x^n area (in.^{2})
12	12.313	119	—	—
10	10.374	84.5	10.75	90.8
8	8.249	53.46	8.625	58.5
6	6.313	31.31	6.625	34.5
4	4.25	14.2	4.5	15.9
3	3.25	8.3	3.5	9.6
2	2.157	3.65	2.375	4.4
1 1/4	1.380	1.5	1.66	2.2
3/4	0.828	0.54	1.05	0.87
1/2 CTS	—	—	0.625	0.31

Table 4-10
Pipe Cross Section—30% and 50% Internal Volume

	Insert Size PE	
Casing Size Steel	For 30% Minimum*	For 50% Minimum
12	6	8
10	6	8
8	4	6
6	3	4
4	2	3
3	1 1/4	2
2	1 1/4	1 1/4
1 1/4	3/4	3/4
3/4	1/2 CTS	1/2 CTS

* After a series of shop tests, it was determined that a minimum of 40% of the internal volume should be occupied to give a margin of safety. This may be achieved by installing a sleeve pipe in addition to the PE pipe. Refer to Chapter 6.

The Right Size

A review of Table 4-10 will show that the desires to use inserts that are two pipe sizes smaller and that also occupy 50% of the volume in the casing do not match. Since difficulty during insertion is a real problem, while some loss of capacity can be tolerated under the worst possible freezing scenario (the loss in capacity is more than compensated by gain due to higher operating pressure), let us select the smaller diameter insert sizes, i.e., those occupying 30% or more of the internal volume of the steel casing pipe. Diameters selected by these guidelines will lend themselves to easy insertion, thus economizing on labor and material costs.

MINIMUM OPERATING PRESSURES

LP System

Since the selection of sizes given on page 57 is such that, upon the water freezing in the annular space in the casing a reduction in the flow area of up to 30% is theoretically possible, we must choose an operating pressure that is high enough to compensate for the squeezing effect. Turning to Table 4-7 and reading the capacity comparison for the selected combination of diameters, it is shown that 105 kPa (15 psi) operating pressure will provide ample capacity in every case. There will be no savings gained by operating at a higher pressure, since the minimum size has already been set by the other considerations mentioned above. This combination of size and pressure provides ample room for increasing the pressure at a later date, thus improving the distribution capacity.

The use of size NPS 6 needs special attention. At this size, PE has a very small price advantage over steel, if directly buried. However, PE can be inserted, and NPS 6 PE insertion can be as much as 40% cheaper than NPS 6 steel directly buried.

To sum up:

1. PE insertion up to and including the NPS 6 size is favored over the direct burial method of steel or plastic. (Based on 1987 prices.)
2. When low-pressure system alone is considered, the new system pressure will need to be raised to 105 kPa (15 psi).

MP System

MP systems operating at 105 kPa (15 psi) need to be raised to 550 kPa (80 psi) to match or exceed the existing capacity. Refer to Tables 4-11 and 4-12 and note the capacity against size combinations marked by asterisks. These combinations meet the 30%-area ratio requirements. Note that a study suggested complete freedom against freezing if this ratio is increased to 40% (Chapter 6). You will notice from Tables 4-11 and 4-12 that the 415 kPa (60 psi) does not meet the capacity requirements, but that 550 kPa (80 psi) provides ample capacity.

MP Capacity Comparison Given Operating Pressure 415 kPa (60 psi)

Merging the existing LP and MP systems into one is desirable as it eliminates one pressure class, which simplifies the operation and maintenance. Merging also eliminates some regulating stations.

Table 4-11
MP Capacity Comparison if Operating Pressure 415 kPa (60 psi)

$$\frac{Q_{415}}{Q_{115}} = \left(\frac{\Delta P_{415}}{\Delta P_{105}}\right)^{0.556} \times \frac{R_{Steel}}{R_{PE}} = \left(\frac{41,132.7}{4,222}\right)^{0.556} \times \frac{R_s}{R_{PE}} = 3.546 \times \frac{R_s}{R_{PE}}$$

Steel Size/PE Insert Size (NPS)	Capacity Ratio	Q_2/Q_1
12/8	3.546×0.112	= 0.801
*12/6	3.546×0.112	= 0.397
10/6	3.546×0.177	= 0.627
8/6	3.546×0.326	= 1.156
*8/4	3.546×0.116	= 0.411
6/4	3.546×0.237	= 0.840
*6/3	3.546×0.121	= 0.429
4/3	3.546×0.348	= 1.234
*4/2	3.546×0.123	= 0.436
3/2	3.546×0.253	= 0.89
*3/1¼	3.546×0.0917	= 0.325
2/1¼	3.546×0.274	= 0.971
1¼/¾	3.546×0.283	= 1.003
*1¼/½ CTS	3.546×0.0719	= 0.255
¾/½ CTS	3.546×0.277	= 0.98

* 40% fill criterion not satisfied. Use a suitable filler or the sleeve pipe to avoid carrier pipe squeeze-off in cold climates.

Table 4-12
MP Capacity Comparison if Operating Pressure 550 kPa (80 psi)

$$\frac{Q_{550}}{Q_{115}} = \left(\frac{\Delta P_{550}}{\Delta P_{105}}\right)^{0.556} \times \frac{R_{Steel}}{R_{PE}} = \left(\frac{68,620.75}{4,222}\right)^{0.556} \times \frac{R_s}{R_{PE}} = 16.25\,\frac{R_s}{R_{PE}}$$

Steel Size/PE Insert Size (NPS)	Capacity Ratio	Q_2/Q_1
12/6	16.25 × 0.112 =	1.48
10/6	16.25 × 0.177 =	2.34
8/4	16.25 × 0.116 =	1.53
6/3	16.25 × 0.121 =	1.6
4/2	16.25 × 0.123 =	1.63
3/1¼	16.25 × 0.0917 =	1.21
2/1¼	16.25 × 0.274 =	4.45
1¼/½ CTS	16.25 × 0.0719 =	1.16
¾/½ CTS	16.25 × 0.277 =	4.5

SYSTEM PRESSURE CHANGES IN OTHER UTILITIES

A telephone survey of North American gas companies who had under-taken similar main replacement programs was conducted by the author in 1986. Table 4-13 shows results of the survey.

In the case of Canadian gas companies, all those who made use of the insertion technique raised the pressure to the maximum limit of PE allowed under the code at the time. This limit used to be 415 kPa (60 psi) for SDR (standard dimension ratio: R = D/t) 11 pipe in class 4 locations. These com-

Table 4-13
Survey of Gas Companies with Mains Replacement Programs

Company	Program Magnitude	Method of Replacement	Operating Pressure Change
1	major—sustained	dead insertion	11 in. to 60 psi
2	insignificant	dead insertion	n/a
3	insignificant	dead insertion	n/a
4	small	dead insertion	11 in. to 60 psi
5	small	—	—
6	moderate—sustained	dead insertion	7 in. to 60 psi
7	major—sustained	live insertion	7 in. to 60 psi (opt'g 35 psi)
8	major—sustained	dead insertion	7 in. to 10 psi
9	major—sustained	dead insertion	7 in. to 15–60 psi
10	minor—sporadic	dead insertion	7 in. to 10 psi

panies state that if they were looking at the design pressure today, they would not hesitate to operate to the present limit of 550 kPa (80 psi).

In the case of U.S. utilities, some are operating up to 410 kPa. Note that U.S. code allows single-cut regulation to 410 kPa only. Since permeation, leakage, and line-break losses are directly related to the line pressure, some companies have qualified their pipe to the maximum code limits, but actually operate at lower pressure. This has given them the flexibility to make full use of the system capacity when required.

If needs are met by 105 kPa, the system is kept at 105 kPa. At least one company in the survey did not raise the system pressure because the replacement mains were not continuous. Although insertion was widely used, it did not encompass any one area completely, as it was done on a piecemeal basis.

Reviewing the practices of North American gas utilities, one may conclude that:

1. It makes sense to use the insertion method (slip lining) in a programmed, systematic fashion.
2. One should take full advantage of the inherent strength of the pipe and operate the distribution system to achieve overall lower costs by plant reduction and improved system efficiency.

MAXIMUM DESIGN PRESSURE

Following is the maximum design pressure for PE piping. System operating pressure should be maintained at a pressure consistent with flow requirements as given in the recommendations.

$$P = \frac{2S}{(R - 1)} \times 10^3 \times 0.32 \qquad \text{(CSA-Z184)}$$

$$S = 8.62 \text{ MPa } (1{,}250 \text{ psi}) \qquad \text{(CSA-B137.4)}$$

$$R = \frac{D}{t}, \text{ given } R = 11$$

$$\therefore \ P = \frac{2 \times 8.62}{11 - 1} \times 10^3 \times 0.32 = 551.68 \text{ kPa } (80 \text{ psi})$$

Note: The use of formula given by the Federal Standard 192.121

$$P = 2S \frac{t}{D - t} \times 0.32$$

gives the same results for SDR 11 pipe.

RECOMMENDATIONS

1. The insertion technique using PE pipe is more economical than the direct burial method of replacement. It is faster, requires less downtime, and is better for public relations. Hence, wherever appropriate, the insertion method should be used.

2. The flow-area reduction resulting from insertion (slip lining) necessitates that the system operating pressure be raised. In the set of conditions described in this chapter, the new pressure for LP should be 105 kPa (15 psi) minimum. Similarly, for MP, operating pressure after insertion should be 550 kPa (80 psi). This assumes that the existing pipes have no spare capacities, loads remain unchanged, and no other system changes take place.

3. The MP and LP systems should be merged, thus eliminating certain regulating equipment and possibly some duplicate mains. The merged systems may require an operating pressure higher than 105 kPa (15 psi). Load data should be updated and network analysis done for LP and MP systems individually and as one, in order to determine:
   a) At what level should the stations be set?
   b) What plant (mains and stations) can be retired?
   c) What new connections, if any, are needed?

4. The new systems should be tested to 770 kPa (110 psi) to qualify them for future operation up to 550 kPa (80 psi).

5. The recommended insert sizes are given in Table 4-7. These are based on two assumptions: that the insert cross-sectional area be a minimum of 40% of the casing area, and that the insert be two sizes smaller than the casing. In most cases it is possible to install just one size down, but it is usually not necessary to do so. Since the 40% rule gives a margin of safety against freeze off, it is recommended that for those size combinations that do not fulfill this requirement, a suitable size sleeve pipe be placed between the PE insert (carrier pipe) and the steel casing.

Sleeve pipe sizes are described in Chapter 13. These sleeves are available through the pipe suppliers.

5
Valving Guidelines

INTRODUCTION

A valving guideline is one of many policy matters that a gas company must decide at the design stage. The development, communication, and enactment of the valving policy will ensure uniformity of design, cost savings, and improved operational efficiency.

Valving guidelines establish working instructions for the design engineer so that he can incorporate an adequate number of valves or similar shut-off devices into his design in order to divide the distribution system into manageable sections for isolation. Although these guidelines are prepared with the mains replacement project in mind, they should be equally applicable and useful in gas distribution system design in general. These guidelines may be applied irrespective of whether or not the insertion technique is used, and whether replacement is by steel pipe or polyethylene pipe.

PURPOSE OF VALVES

The primary purpose of a valve (Figure 5-1) is to shut off the gas flow in an emergency, so that the gas escape can be controlled and stopped with a minimum loss of gas and damage to property and personnel.

Valves also serve another purpose; they provide a convenient means of tying, routine maintenance, and control and isolation of gas systems for various operational reasons, for example, isolating the heat areas.

If valves are accessible, installed in proper locations, and maintained regularly, they are the best method for shutting off the gas flow in emergencies, especially if other methods of flow control are not locally available.

Figure 5-1. Valves installed at strategic locations will facilitate future operation and maintenance of the pipeline system.

BASIC CONSIDERATIONS

The operational convenience that valves provide has to be judged against the installation cost plus the regular maintenance requirements associated with the valves. The desire to have an abundant availability of valves to isolate the distribution system in small segments for repairs, while at the same time causing no discomfort to customers, is laudable. But it must be weighed against the associated problems, namely, additional capital cost, potential leaks, and high maintenance requirements.

Most underground valves cannot be operated in an emergency because of various problems, such as the valve cannot be found, the valve box is filled with silt or ice, the valve is not operable, or an important customer downstream from the valve cannot be without gas.

Leaking valves are generally not a serious hazard, as the leaks are usually in the glands and the gas is able to vent through the valve box. A good maintenance program can reduce or eliminate the leak problem. Operability and accessibility can only be achieved through a program of identification and regular maintenance. This costs money, but it is hardly worthwhile to install valves in the first place if they are not properly maintained, as they must perform satisfactorily in an emergency. Some valves require less maintenance than others.

The ideal situation is to have the least number of inexpensive, maintenance-free valves, located in strategic places so that the gas system can be isolated in a small number of meters. The bigger the isolated area, the greater the potential gas loss, relight time, and customer-service manpower needs.

OPTIONS FOR SHUT OFF

Shut off can be achieved in a number of ways, such as by closing certain valves or by the pipe-squeezing or line-stopper methods. Since both the line-stopper and pipe-squeezing methods involve exposing the line, which delays the shut off (particularly in the winter), they are not the best options for emergency shut offs. Also, the PE insertion technique limits the accessibility of the pipe for squeezing. Further, the pinch off will later require exposing the main to install a clamp over the pinched pipe. Accessibility of pipe is similarly restricted for line-stopping operations, due to the encasement of the PE pipe into the steel casing pipe.

Valving is the most suitable choice for shutting off gas in an emergency. This leads us to the question of the type and quantity of valves you should install.

TYPES OF VALVES FOR PE SYSTEMS

Ideally the underground valve should be:

1. The same material as the pipe.
2. Non-lubricated type, requiring no maintenance.
3. Full port.
4. Operable from aboveground and available with water-tight high-head extension.
5. Designed for low-pressure drop.
6. Capable of a 1/4-turn shut off.
7. Proven to be reliable in similar service conditions.
8. Inexpensive, with a short delivery time.
9. Capable of meeting the requirements of API, ANSI, and MSS (CSA in Canada).

CODE REQUIREMENTS

United States DOT requirements are as follows:

192.181 Federal Standard

192.181 Distribution line valves (11-12-70)

(a) Each high-pressure distribution system must have valves spaced so as to reduce the time to shut down a section of main in an emergency. The valve spacing is determined by the operating pressure, the size of the mains, and the local physical conditions.
(b) Each regulator station controlling the flow or pressure of gas in a distribution system must have a valve installed on the inlet piping at a distance from the regulator station sufficient to permit the operation of the valve during an emergency that might preclude access to the station.
(c) Each valve on a main installed for operating or emergency purposes must comply with the following:
    (1) The valve must be placed in a readily accessible location so as to facilitate its operation in an emergency.
    (2) The operating stem or mechanism must be readily accessible.
    (3) If the valve is installed in a buried box or enclosure, the box or enclosure must be installed so as to avoid transmitting external loads to the main.

Clause 5.10.1.2 and 5.10.2.2.1 of CSA Standard Z184 M86 give minimum Canadian standards requirements for valving on distribution systems. These clauses are repeated here for reference.

5.10.1.2 Distribution Systems

Valves in distribution lines, for operating or emergency purposes, shall be spaced as follows:

(a) For high pressure distribution systems, valves shall be installed in accessible locations, in order to reduce the time required to shut down a section of the line in an emergency. In determining the spacing of the valves, consideration shall be given to the operating pressure and size of the distribution lines and local physical conditions, as well as the number and type of consumers that might be affected by a shutdown.

(b) For low pressure distribution systems, valves shall not be required, except as specified in Clause 5.10.2.2.1.

5.10.2.2 Distribution System Valves

5.10.2.2.1. For distribution systems, valves shall be installed to provide for shutting off the flow to each regulator station controlling the flow or pressure of gas in the distribution system. The distance between the valve and regulators shall be sufficient to permit the operation of the valves during an emergency, such as a large gas leak or a fire in the station.

VALVE SPACING

Mains replacement projects generally take place in the older parts of the cities, i.e., in class 3 and 4 areas where high-rise buildings and commercial customers are prevalent. These are comparatively larger loads and some customers have more than one use for gas, such as heating and cooking. Therefore, a higher priority should be given to these customers. Gas outages in these areas have the potential for higher revenue loss. Construction activity and the concentration of utilities is also greater in class 3 and 4 areas, hence the gas pipe in higher-density areas is more damage-prone. Also, the gas leaks in these areas are potentially more hazardous. If the insertion method is used, the plastic pipe is encased in a metal pipe and hence not readily accessible for pinching off. In view of the above, the following valve spacings are recommended for the distribution mains:

Class 4 Areas—isolation of 4 city blocks or 100 customers, whichever is
             less.
Class 3 Areas—500 customers.
Class 1 and 2—500 customers.

The recommended valving frequency is one adopted by some gas utilities in Canada, and is given here as a general guide. Each company must review its system requirements, customer density, and service area before deciding on valve spacing.

6
Scouring and Freezing

INTRODUCTION

Gas mains and services replacement can be done by either the conventional method of direct burial, i.e., trenching and laying new pipe alongside the old one, or by the insertion technique, in which case the new pipe (referred to as the carrier pipe or the PE pipe) is inserted into the old pipe (hereinafter referred to as the casing or the steel pipe). In the insertion technique, considerably less excavation and therefore less inconvenience to traffic is involved, resulting in costs that are up to 40% less than for the same work if done by the direct burial method.

However, during the process of inserting the plastic pipe into the metal pipe (steel or cast iron), the plastic pipe can be damaged. In cold climates, the squeezing of plastic pipe by ice formations within the metal pipe is also a concern. To protect the plastic pipe from external damage during insertion and subsequently from further damage due to expansion, contraction, and ice formations squeezing the plastic pipe, a gas company may choose to install sleeve pipes between the carrier pipe (PE pipe) and the steel mains that are to be abandoned.

SCOURING

Since the inserted PE pipe and the casing have standard pipe dimensions, you can only insert a PE pipe that is at least one size smaller than that of the casing. Damage to the exterior surface of the PE pipe is more severe and insertion more difficult as the void between the outside diameter of PE pipe and the inside diameter of the casing becomes smaller. It has been found that for ease of insertion, the PE pipe should at least be two sizes smaller than the steel pipe. Even with this two-size-down rule, PE pipe gets badly scoured in some insertions. The internal surface of the steel pipe determines

the degree of damage. Weld icicles, bends, projecting service tees, rust parti-
cles, debris in the casing, and mechanical joints all add to the severity of the
damage. Canadian codes allow scratches up to 10% deep. In some instances, this damage is more than 10% of the wall thickness. As the location
of the sources of damage cannot be easily determined, it is not practical or
economical to find and remove these troubling sources. Running a poly pig
through at the start of the job is a standard method. It helps to remove the
debris. But pigging does not knock off weld spatters or rough weld beads,
nor does it straighten the misaligned components of mechanical joints, such
as Dresser couplings, etc.

Solutions to Scouring

A number of solutions to the scratch problems exist, including:

1. Ignoring the problem.
2. Using thicker-walled PE pipe to compensate for the depth of scratches.
3. Installing Thinsulators over the PE pipe before insertion.
4. Inserting a plastic sleeve in the steel pipe before inserting the PE carrier pipe, or inserting the sleeve and PE pipe together.

Ignoring damage to the pipe beyond 10% wall thickness will be a non-
compliance of Canadian codes. Use of thicker-walled pipe is a good solution, and is the cheapest of the alternatives above. The disadvantages of
thicker-walled pipe are that flow capacity is reduced by about 15% when
you switch from SDR 11 to SDR 9. (SDR is defined as the outside diameter
of pipe divided by its wall thickness.) Also, the drilling of service tees becomes difficult, particularly so in larger sizes.

The Thinsulator option is an expensive one. A plastic sleeve installed between the steel pipe and the PE carrier pipe will provide complete protection against protruding objects. It provides a smooth plastic surface on
which the carrier pipe can slide easily during the installation. In later years,
it will provide a smooth bed on which the carrier pipe will move back and
forth several times annually, as the differential expansion and contraction
takes place between the PE pipe and the steel casing due to temperature
changes.

FREEZING

In an inserted main, the void left between the PE carrier pipe and the
steel pipe can accommodate ground water. The water will find its way into

this void through corrosion holes, leaking fittings, and ends where the casing is cut for inserting the carrier pipe. In cold climates, the frost depth may be as deep or deeper than the casing. Any water trapped in the steel casing, will, due to the cold ground temperature, freeze and become ice. Since ice has a higher specific volume than water, it needs more room. Thus, the ice pushes the plastic pipe inward and the casing outward. Since plastic material is relatively weak, the ice can easily push the plastic pipe inward and may completely flatten it.

It has been found that areas where the water table is high or that have pockets of clayish soil impeding the percolation of rain water to lower ground are more susceptible to ice formation under sub-zero conditions. Pockets of clayish soil and frost enveloping the casing are ideal conditions for trapping and freezing the water in the casing.

North American gas companies who operate under similar soil, ground-water level, and sub-zero winter conditions have experienced this freezing phenomenon, and have taken steps to forestall or minimize the effects of ice formation. A survey of some of the North American utilities is given in Table 6-1.

If damage to the carrier pipe is expected, installation of a plastic sleeve is the best answer. A sleeve extending beyond the steel-pipe ends will also provide shear protection to the carrier pipe.

Criteria for the Use of a Sleeve

Under certain conditions a reduction in the PE pipe cross-sectional area will occur. Under suitable conditions the gas flow may stop completely. Suitable conditions are:

1. The presence of water in the void between the PE pipe and the steel pipe.

2. Frost or some other source lowering the temperature of water to below its freezing point.

3. A void large enough to allow a critical amount of water.

All three conditions must be met to cause flow stoppage by flattening the plastic pipe.

Control over the ground water and its ingress into the steel casing is impractical. Similarly, frost penetration is a climatic condition beyond the gas

Table 6-1
1986 Survey of North American Utility Companies on Insertion

Company	Insertion Program	Frost Depth	Mains Cover	Mains Material	Freezing Problem	Remarks
1.	Mains-Minor	2 ft	3 ft	CI (Cast iron)	—	Pipe is pushed, not pulled. Anchors at ends.
2.	Pilot Proj.			CI	—	A few services and mains have been done.
3.	Minor			Steel	—	A few services done.
4.	Pilot Proj.			CI	—	—
5.	Pilot Proj.	.75 ft	2 ft	CI	—	Standards/sizing per AGA 1977 Operating Section Proceedings, i.e., PE to be 2 sizes smaller than steel casing.
6.	2 miles of various sizes steel	3 ft	2.5 ft	Steel	One	Insertion sizing is done as above. A service line that was not sized per Co. standards froze off. This was a 1/2-in. CTS copper inserted into 11/4-in. steel.
7.	10 miles + 1,000 serv.	4.5 ft	3 ft	Steel	—	PE size is one size smaller than steel casing. This standard was followed from day one. No freezing problem. Some scratches are deeper than 10%, but are left in for trial purposes.
8.	Substantial	4 ft	3 ft	CI, Steel	X	One catastrophical failure of a 12-in. steel main (60 psi) encased in a 24-in. steel main occurred. Insertion program was halted and company procedures were changed. Freezing problem is now prevented by following one of the techniques listed on the next page.

(Table 6-1 continued on next page)

Table 6-1 continued

Company	Insertion Program	Frost Depth	Mains Cover	Mains Material	Freezing Problem	Remarks
8. (con't)						(a) Direct burial (b) Fill void with a non-shrinking grout. (c) Fill casing with old PE pipe. (d) Polyurethane foam injection into casing. (e) Oversizing the PE carrier pipe.
9.	Substantial				X	After 6 services out of the first 300 froze, casing/carrier pipe sizes were chosen such that water volume is limited.
10.	One insertion, 2 blocks long where 4-in. PE inserted into 6-in. steel main			Steel	—	—
11.	Service insertion only	4 ft	3 ft	CI	X	Some 1/2-in. CTS services inserted into 1 1/4-in. pipe froze. They now use the largest PE pipe that can go inside the steel pipe or use the direct burial method.
12.	Substantial	2 ft	3 ft	CI	—	Steel system is approximately 10%, the remainder is cast iron. Damage to PE during insertion nor freezing has ever been experienced. Water table is low, i.e., below 10 ft and frost depth is only 2 ft.
13.	Substantial	3.5 ft	3.5 ft	CI	—	Freezing never a problem. This company does not use 1/2-in. CTS at all. PE not inserted in steel, only in cast iron. No PE by direct burial in class 4.

company's control. However, the void can be controlled. This may be done by any of the following methods:

1. Filling the void with a grout.
2. Filling the void with a foam.
3. Oversizing the PE pipe.
4. Double insertion, i.e., the use of a sleeve.

The grouting method was found to be impractical for small sizes. It is also expensive at approximately $10–$15 per meter. Filling the void by injecting foam is another alternative. The polyurethane foams generally have a short setting time, not allowing enough time for the foam to flow in the casing. Providing there is enough clearance (150 mm), this shortfall can be overcome by using moving injection nozzles for foaming or by blowing small foam beads in the annular space with air. Oversizing the PE pipe may seem a good alternative, but it will cause more pipe scratches and pipe hang-ups during insertion.

British Gas has developed a tight lining technique called swage lining, in which all weld icicles are shaved off the casing using a steel pig, and every internal obstruction is physically removed. Plastic pipe of the same size as the casing is passed through a swaging machine on site, reducing the outside diameter of the casing. Reduced diameter PE pipe is inserted in the casing, which, a few hours after installation, relaxes, expands, and clings tightly to the casing, leaving no room for liquids to accumulate. The cost of tight lining is higher than that of slip lining and, depending on the size and location, may not have any advantage over direct burial. However, it should be viewed as an alternative where higher flows are required.

Double insertion, i.e., the sleeve, provides the best alternative again. It is inexpensive and it facilitates insertion, thus reducing the construction time. Simulated freeze testing has shown that the deformation of the PE carrier pipe caused by the ice formation in the casing is non-uniform.

Example 6-1

| Size NPS | Steel Pipe | | | PE Pipe | | Void |
	ID in.	Internal X^n Area in.^{2}	OD in.	Total X^n Area in.^{2}		X^n Area in.^{2}
4	4.25	14.2	—	—		14.2–4.4
2	—	—	2.375	4.4		= 9.8

Flow area of NPS 2 PE SDR 11 pipe is $\pi/4 \times (2.375 - 0.432)^2 = 2.96$ in.^{2}
Consider a unit length.
Increase in area of water upon icing = $0.1 \times 9.8 = 0.98$ in.^{2}
Increase in X^n area of ice = Decrease in X^n area of flow

New flow area = $2.96 - 0.98 = 1.98$ in.^{2}

$$\text{Percentage reduction} = \frac{\text{original area} - \text{new area}}{\text{original area}} = \frac{2.96 - 1.98}{2.96}$$
$$= 33\%$$

The calculations show that a loss of the cross-sectional area of the PE pipe will occur. This should be a maximum of 33% if the void was completely full of water and all the water froze uniformly. Shop tests showed that the squeezing was not uniform. In some sections, the plastic pipe remained almost circular, while in others the reduction was more than 33%. (See Figures 6-1 through 6-5.)

Figure 6-1. Ice formation in 60-mm steel casing, causing 15-mm PE tube to deform at 2 spots in 3-m test length.

Figure 6-2. Ice formation in 42-mm steel casing causing tubing to deform throughout the 3-m test length.

Figure 6-3. Ice formation in 60-mm steel casing causing 26-mm PE pipe to deform throughout its entire frozen length.

Figure 6-4. Ice formation in 60-mm steel casing causing 42-mm PE pipe to deform at 2 spots in 3-m test length.

Figure 6-5. Ice formation in 114-mm steel casing causing spot deformation on 60-mm PE pipe.

It is theorized that the squeezing of PE pipe starts at the weakest point. As the water turns into slush, the distribution of the squeezing pressure becomes non-uniform. Areas where slush is still fairly fluid relieve the hydraulic pressure by expanding longitudinally along the pipe, hence they bear little deformation. On the other hand, a section of pipe that still has unfrozen water/slush trapped between two completely frozen sections will undergo a disproportional deformation. In an active gas-distribution system, the flowing gas will tend to distribute the heat along the pipe and will make freezing, and hence squeezing, more uniform.

Since the steadying effect of the gas flow cannot be determined for all pipe locations, it may be discounted, thus erring on the side of safety. When the ratio of cross-sectional area of PE pipe to the internal cross-sectional area of casing remained above 40%, spot squeezing did not occur. This ratio may be termed as the critical ratio. Using this criterion of 40%, suitable carrier-pipe-to-steel-pipe combinations may be chosen. When the carrier pipe occupies more than 40% of the casing volume, sleeving is not required unless scouring is a problem.

CONCLUSIONS

The sizing of PE pipe for the insertion project has to meet the following requirements:

1. **Load Considerations.** A review of Chapter 4 shows that in order to serve the same load, the inserted PE pipe may be two pipe sizes smaller than the original steel pipe, providing the operating pressure is raised from 1.72 kPa to 105 kPa.

2. **Insertion Ease.** The experience of North American utilities indicates that the insertion of the PE pipe in a steel or cast-iron pipe is trouble-free as long as the PE pipe size is at least two sizes smaller than the size of the casing. Tighter combinations invite more damage to the PE pipe.

3. **Freezing Considerations.** An analysis of the freezing phenomenon shows that complete closure of PE pipe can be ruled out with certainty as long as the cross-sectional area of the PE pipe is at least 40% of the internal cross-sectional area of the casing. This 40% rule allows two sizes down in some cases, but requires one size down in other size combinations. The term "two sizes down" is used for simplicity. The actual suggested sizes have some exceptions. Table 6-2 shows the steel/PE sizes and their respective ratios.

   We can choose to have two sizes down in every case and bring the ratio to above the 40% mark for those sizes requiring it by providing a sleeve pipe in between the steel and the PE carrier pipe. Suitable sleeves have been designed that allow proper clearances and volumes. Refer to Table 6-3 for sleeve sizing.

4. **Scouring Problem.** The scouring problem can be countered either by using thicker wall pipe or by the use of a sleeve.

Table 6-2
Ratios of Cross-Sectional Areas of PE Pipe to Inside Cross-Sectional Area of Steel Pipe

Steel Size NPS	PE Size NPS	Ratio %
¾	½ CTS	57
1¼	¾	58
2	1¼	60.2
3	*1¼	26.5
4	*2	30.0
6	*3	30.6
8	*4	29.7
10	6	40
12	*6	29

NOTES
1. Distribution steel line pipe considered.
2. SDR 11 PE Pipe considered.
* Use of sleeve is strongly recommended.

Table 6-3
Sleeve Sizes

OD (mm)	Steel Pipe Size NPS (in.)	ID (mm)	PE Carrier Pipe Size OD (mm)	Sleeve OD (mm)	WT (mm)
33.4	1	26.6	15.9	21	1.5
42.2	1¼	35.1	26.7	31	1.5
60.3	2	52.5	42.2	47	1.5
88.9	3	82.55	42.2	60.3	3.8
114.3	4	107.95	60.3	88.9	3.18
168.3	6	160.3	88.9	136.65	4.5
219.1	8	209.5	114.3	190.0	6.0
			168.3	190.0	6.0
273.1	10	263.4	168.3	219.1	3.18
			219.1	250.0	5.7
323.9	12	312.67	168.3	250.0	10.7
			219.1	250.0	5.7
304.8	12 in. OD	292.1	168.3	250.0	5.7

Note: For steel pipe sizes 60.3 mm (2 in.) and smaller, sleeving for the entire length of steel pipe is not required. For these sizes, use sleeve pups at ends only.

RECOMMENDATIONS

When you consider all aspects of freezing, scouring, ease of insertion, shear protection, cost, and capacity, the use of a plastic sleeve clearly provides the best solution, as it facilitates insertion, provides freedom from scouring damage, and protection against freezing.

7
Budget Costs

INTRODUCTION

Budget preparation requires an assessment of what is to be done, how it is to be done, when, by whom, and at what cost. Answering what is to be done requires defining the scope of the project. Chapter 2 gives some of the information on how data can be collected and assembled to define the scope of work. Maps, records, and leak histories must be studied to assess the extent of the work. Once information on lengths, diameters, pressures, location, depths, pipe materials, loads, meter locations, regulating stations, etc., is complete, the next step is to estimate how much insertion and how much direct burial is involved. Operating pressures and company practices will determine what materials and equipment will be needed. In a large distribution renewal project, 100% insertion is not practical, as considerations of capacity, routing, depth of cover, and obstructions necessitate some direct burial. The direct-burial percentage may be as high as 20%. The use of company labor versus contract labor must be taken into account. The annual budget will depend on how much work is done in a particular year.

COST COMPONENTS

A distribution renewal project using the dead insertion method may entail the following cost components:

1. **Mains**—80% insertion, 20% direct burial is a good assumption.

2. **Services, including risers**—Percentage of insertion may be assumed to be 80%.

3. **Meter relocations from inside to outside and the installation of outside regulators.** This component of cost is not a necessary part of distribution system renewal. To facilitate meter reading, some companies may find it an opportune time to relocate the meters from inside to outside. System replacement by insertion necessitates raising the operating pressure, which in turn means installing regulators on the services. The relief from the regulator must vent outside the building. The relocation of a meter from inside to outside also means some make-up piping work inside the premises.

4. **Temporary gas supply.** Some customers, particularly commercial and industrial customers, may require a temporary gas supply while work on their main/service line is being done. It may be assumed that no more than 2% of the customers will require temporary gas.

5. **Regulating stations abandonment/modification.** A change in system pressure will require changes to the equipment at the regulating station. With realignment and resizing of mains, some stations may become redundant.

If it is possible to delay preparing the budget until after the network analysis is complete and the mains and services have been sized, very accurate budget numbers can be prepared. If accurate design is not available before the budget preparation, budgeting can still proceed by making certain simplifying assumptions. These are:

1. That all pipe defined as such within the project shall be required, hence the total length will remain the same.
2. System capacity is to remain the same, hence equivalent pipe sizes can be used.
3. Construction will be by insertion (80%) and direct burial (20%).
4. Existing routing and ground surfaces shall be maintained.

Once sizing is complete, budget figures should be updated.

COST BREAKDOWN

Mains (A)

Assume contract work. Assume 80% insertion, 20% direct burial. Replace existing steel main by PE pipe.

Replacement Sizes	Cost
NPS 2	Length × 0.8 × (unit insertion contract price + material cost)
	+
	Length × 0.2 × (unit direct burial contract price + material cost)
NPS 4, 6, etc.	As above

Services (B)

Assume an average length of 25 m. Assuming an average flow rate for residential services, pipe size can be determined at 26 mm PE.

Cost = Quantity × 0.8 (unit insertion contract price + material cost)
       Quantity × 0.2 (unit direct burial contract price + material cost)

Meters (C)

Install regulator, move meter, reconnect to customer's piping.

Cost = No. of customers × (unit labor price + material price)

Temporary Gas Supply (D)

Assume 2% of customers will require temporary gas supplies. Normal supply is restored within 8 hours.

Cost = 2% of affected customers × estimated (hook up + gas) cost based on 8-hour consumption

Regulating Stations (E)

1. Stations to be abandoned = no. of abandonments × abandonment cost
2. Stations to be revamped = no. of revamps × revamp cost
3. New stations = no. of new stations × new station cost

COST SUMMARY

1.	Mains		$A
2.	Services		B
3.	Meters/Regulator		C
4.	Temporary Gas Supply		D
5.	Regulating Stations		E
6.	Total Construction Cost		= $F
7.	Field Research	1 % of F	
8.	Engineering	10 % of F	
9.	Drafting	2 % of F	
10.	Field Inspection and Construction Supervision	5 % of F	
11.	Material Handling	2.5 % of F	
12.	Subtotal EPC (engineering, procurement, construction)	.205 F + F	= 1.205 F
13.	Administration and Overheads	5 % EPC	= 0.060 F
14.	Subtotal		= 1.265 F
15.	Contingencies 10% of F		0.1 F
16.	Total Project Budget		= $1.365 F

ANNUAL EXPENDITURE

Let us assume that the total project budget is $1.365 F, and that you are gas utility "Y" who has to replace 200 km of distribution mains and 18,000 services of various sizes in an urban area. This work will require careful planning, liaisons with government authorities, contractors, and customers. In order to keep traffic disruptions, street closures, gas outages, and complaints under control, work must be carefully planned to be carried out over a period of time. A sample calculation to determine this period follows:

1. Based on construction season of May to November, 5 days/week work, total annual rain-free working days during summer construction period = 100.
2. Number of crews = 3. Each crew disrupting up to 6 blocks simultaneously, i.e., subcrews working on excavation, tapping, insertion of main and services, house piping, backfilling, paving, clean up, and moving. This may be considered the maximum acceptable level of disruption, i.e., $3 \times 6 = 18$ city blocks at any one time.
3. Number of customers shut down/crew on insertion day = 30. Since the actual turn off is only once a week, the average daily turn off per

crew = 6. Because crews cannot be perfectly synchronized, shutdowns may coincide, and in the worst case up to $30 \times 3 = 90$ customers could be without gas on any one day. Your company may consider this number to be the maximum for any one location at a time.

The larger the customer shutdown, the larger the customer's temporary gas requirement, relighting needs, and complaints. Complaints can be minimized with an aggressive customer-relations information effort and by providing a well-informed telephone answering service. Based on the calculations given below for the project duration, an annual expenditure of 10% of the total project estimate (in constant dollars) is required for each of the next ten years.

PROJECT DURATION

Number of customers shut down/day/crew × number of crews × number of working days/year × number of years = Total number of customers = 18,000

Therefore $6 \times 3 \times 100 \times Y = 18,000$
$$Y = 10 \text{ years}$$

We calculated that the total project cost estimate is $1.365 F. Hence, the first year budget amount should be $0.1365 F. If work is done in equal portions, is spread over a 10-year span, and an annual inflation rate of 5% is assumed, the project's actual dollar requirement will be as follows:

Year	$(F)
1	0.1365
2	0.1430
3	0.1500
4	0.1580
5	0.1660
6	0.1740
7	0.1830
8	0.1920
9	0.2020
10	0.2120
Total	1.7165

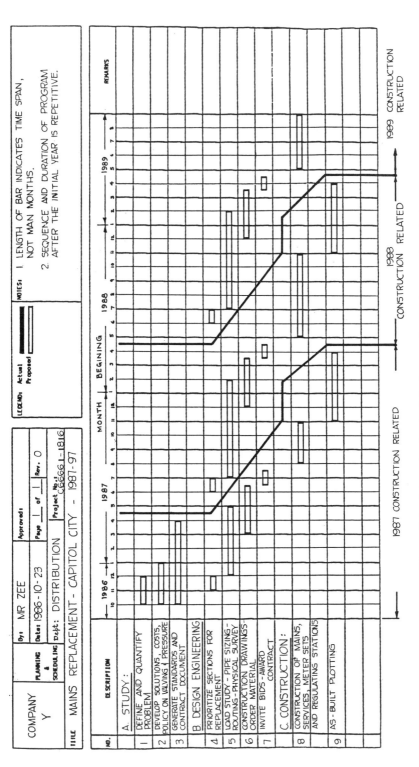

Figure 7-1. A bar chart schedule for a mains replacement project.

A Bar Chart Schedule

A bar chart schedule lists all project activities for pipeline replacement in one column, and the desired start and completion dates opposite them. The activities should preferably be listed in the sequence in which they should occur.

Figure 7-1 shows an actual example of the use of a bar chart for an insertion project undertaken by one company.

8
Deciding Future Operating Pressure

INTRODUCTION

In Chapter 4, equivalent pipe sizes were calculated for the replacement of the steel pipe with new plastic pipe. The new operating pressure may be chosen from 1.72 kPa (¼ psi) to the maximum allowed for the PE pipe. Pressures of 105 kPa (15 psi), 210 kPa (30 psi), and 550 kPa (80 psi) are considered here.

In order to decide the future operating pressure, it is necessary to explore all practical options available and their relevant merits and costs. Generally speaking, the higher the system pressure, the greater the capacity or the smaller the size of pipe that can be used to achieve the same flow rate. However, as you raise the pressure you also raise the potential risk in the event of pipe failure. Various options discussed in this chapter should be considered.

PARTICULAR CONSIDERATIONS

Operating Pressure 105 kPa (15 psi)

1. Single-cut service regulators will be required.
2. Merging of LP and MP (105 kPa) systems takes place, thus simplifying pressure classes by one.
3. District regulating stations require modification to give new outlet pressure.
4. LP commercial and industrial sales stations require modifications.
5. Maintenance operators require retraining to work on higher pressure mains.

Operating Pressure 210 kPa (30 psi)

Items 1, 3, 4, and 5 from above apply. Additionally, this option eliminates two pressure classes, i.e., LP; and MP-115 kPa and MP-210 kPa are merged into one.

Operating Pressure 550 kPa (80 psi)
---

1. Single-cut service regulators will be required.
   Note: In the U.S., double-cut regulation is mandatory for service lines operating above 410 kPa (60 psi). See Federal Standard 192.197.
2. All distribution pressure classes, i.e., IP, MP, and LP merge into one, simplifying mains and regulating station operation. Many mains and all district regulating stations can be retired. Only the city gate stations feeding IP (550 kPa) gas into the project area need to operate.
3. Sales stations will require modifications, i.e., pressure regulation upstream of the existing meters will be necessary.
4. Maintenance operators will require retraining to work on higher-pressure mains.
5. Mueller low-pressure drilling and flow-stopping equipment has a range of up to 410 kPa (60 psi). Hence, for operations on steel sections, high-pressure Mueller or Williamson equipment and fittings will be required. However, it should be noted that the equipment needs for the drilling and flow stopping will not change substantially, as the new system pipe is substantially plastic, which can be flow-stopped by squeezing.
   Note: Any operating pressure change will require some change of inventory at the warehouse.

DECISION-MAKING CHART

After a detailed analysis of operation and cost for each scenario is worked out, a score chart as shown in Table 8-1 may be prepared. This chart takes into account investment and operating concerns in a numeric form. All important considerations are listed and allotted a weight factor that is arrived at by consensus of the management team. Then each pressure option is given a rating from 1 to 10 for each important consideration, termed as "quality." The detailed analysis carried out earlier must be translated into simple terms under the heading of "quality" in the score chart.

Table 8-1
Score Chart for Operating Pressures

Weight Factor	Quality	Pressure kPa		
		105	210	550
6	Low capital cost	8	9	10
		48	54	60
8	Low O & M cost	10	10	10
		80	80	80
10	Minimize gas escape at break	10	8	4
		100	80	40
2	Minimize freezing (squeeze-off effect)*	10	8	7
		20	16	14
2	Minimize number of different pressures	10	5	10
		20	10	20
	Totals	268	240	214

* Standard pipe dimensions are such that when a smaller size pipe is inserted into a larger pipe, a void is left in between which collects ground water. Winter frost may cause this water to freeze. Water when freezing requires more room, and hence it deforms the plastic pipe inward.

Table 8-1 suggests that 105 kPa (15 psi) operating pressure is the most favorable pressure. The chart is produced to show how ideas and preferences can be evaluated in a numeric form to make the decision making easier.

Your company may have a different set of circumstances and values, and your management may use a similar technique to arrive at a different decision. The example shown above is an actual case; accordingly, a gas utility decided that the system operating pressure in the project area, which includes the downtown central core and related high-density residential districts, should, after insertion, be raised to 105 kPa (15 psi) from the then low pressure of 1.72 kPa (1/4 psi).

9
Critical Path Analysis

INTRODUCTION

In this chapter, an introduction to the critical path method (CPM) is presented, followed by illustrations of application of CPM to a distribution renewal project. Two examples are given. Table 9-1 covers the project period from the time the customers are given notice of the gas company's intention to do the work until the time all replacement work is finished and the cleanup of the work site is completed.

CRITICAL PATH METHOD (CPM)

Critical path analysis is a method of logically listing all project tasks, their duration, sequence, and interrelationships to schedule events to occur during a certain time and at a predetermined cost, thus enabling the management to complete the project in time and/or with the least expenditure.

Critical path analysis forces the project team to think logically and plan thoroughly before the job starts. When the critical path diagram has been drawn in its final form, it serves as a master plan that can be updated or amended as the work proceeds. The critical path is the path through the CP diagram (Figure 9-1) on which any delay will immediately lengthen the completion time. A step-by-step method for creating a simple CP diagram follows. (For a more rigorous treatment of this subject, the reader is advised to refer to a textbook.) Also for project planning purposes, PC software is readily available which can be used not only to carry out lengthy calculations but also to quickly produce multiple iterations of critical path diagrams. However, before PC software can be used, it is important to know the terminology and the basics of CPM, explained here.

Table 9-1
Replacement Tasks—Typical Residential Block

Event Number	Prerequisite	Event	Estimated Duration	Description
1.		Info package	5 days	Warn residents of impending construction
2.		Service orders search	10 days	Photocopy old service orders in project
3.		W.O. (meter sets)	10 days	Customer service work inside house
4.	1, 2, 3	Survey (customer)	5 days	New meter location, safety hazards
5.	4	Extend house piping	10 days	New pipe to outside meter location
6.		Depth survey	4 days	Walk all lines with a 810 locater
7.	6	W.O. (mains & services)	20 days	Contractors W.O.
8.	7	Traffic & customer outage notices	5 days	Deliver notices to customers
9.	7	Excavation permit from city	1 day	City permits for excavating
10.	8, 9	Excavation of mains & services	(1 day) 8 hr.	Dig out old main
11.	10	Install stoppers	(1 day) 8 hr.	Weld on fittings
12.	10	Run new services, if any	(1 day) 8 hr.	Dig in new service lines
13.	10	Assemble services	(1 day) 8 hr.	Butt weld, anodeless riser, etc.
14.	5, 11, 12, 13	Turn off customers	½ hr.	At Luboseal or curb stop
15.	14	Bagging off main & purge	1½ hr.	Stop flow of gas
16.	14	Meter disconnect	2 hr.	Remove old meter
17.	15	Cut out service taps	1½ hr.	Remove old main at service tee location
18.	17	Install main & test	2 hr.	Insert PE pipe in old main
19.	14	Install services	1 hr.	Insert PE services
20.	18	Final main tie-in, purge	1 hr.	Tie-in new MP supply
21.	19, 20	Attach tee & test services	2 hr.	Fuse service tee
22.	21	Tap service tee & purge	1 hr.	Drill out tee, purge out air
23.	16, 22	Set meters, relight & safety survey	1 hr.	Inspect interior pipe and new meter
24.	23	Backfill	(1 day) 8 hr.	Excavation filled, surplus removed
25.	24	Cleanup	(1 day) 8 hr.	Backfill at house & tamp, cleanup, restore

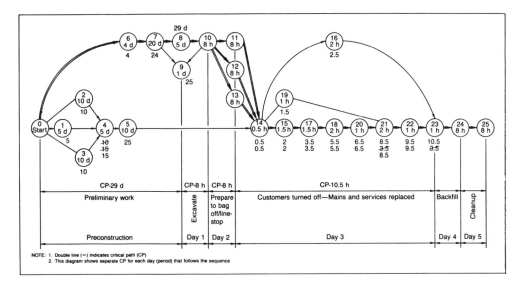

Figure 9-1. Critical path diagram for a gas distribution system renewal project.

Drawing a Critical Path Diagram

1. → Use to join tasks or events together.
2. ○ Use to show termination of tasks or events.
3. List all events.
4. Sort out all events by the sequence in which they occur.
5. Estimate the time likely to be taken by each event. Mark this time in the appropriate circle.
6. When an event can only be started after more than one event must finish, the earliest starting time for the next event will be the longest time required to finish all preceding events.
7. To calculate the earliest event time, add to the earliest time of each immediately preceding event the duration of the activity which connects it, and select the highest of the values obtained.
8. To calculate the latest event time, subtract the duration of the task which connects it from the latest time of each immediately succeeding event, and select the lowest of the values obtained.

Example 9-1

Given that the project requirements are:

Event 2 must follow event 1, and requires 0.5 day to complete.
Event 3 must follow event 1, and requires 3 days to complete.
Event 5 must follow event 3, and requires 8 days to complete.
Event 4 must follow event 2, and requires 1 day to complete.
Event 4 must follow event 3, and requires 1 day to complete.
Event 6 must follow event 4, and requires 1.5 days to complete.
Event 6 must follow event 5, and requires 1.5 days to complete.
Event 7 must follow event 6, and requires 8 days to complete.
Event 7 must follow event 5, and requires 8 days to complete.

Draw a diagram showing the earliest event time, the latest event time and the critical path.

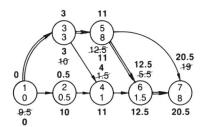

Earliest event time is given in heavy numerals above the event number.
Latest event time is given in heavy numerals below the event number.
Critical path—where earliest and latest event times are the same, i.e., with zero float.

Event	Earliest Completion	Latest Completion	Float
1	0	0	0
2	0.5	10	9.5
3	3	3	0
4	4	11	7
5	11	11	0
6	12.5	12.5	0
7	20.5	20.5	0

Total float is determined as the difference between the latest completion time and the earliest completion time of an event. Events that are on the critical path have no float time. Hence, events 1, 3, 5, 6, and 7 are on the critical path. The critical path is shown by double lines.

Example 9-2

A distribution renewal project was analyzed using the critical path method. From an organizational and planning point of view, it was important to analyze how long the customers would be without gas and how long the customers' yards and the streets/lanes will be disrupted, so that proper coordination of garbage pick-up, deliveries, and manpower can be organized. It is interesting to note from the CP diagram that day 3 (the day on which the customers' gas is shut off) is a 10½-hour working day. Hence, if 5 crews are working simultaneously, one behind the other (one for each job from day 1 to day 5), the crew working on day 3 will be required to work a 10½-hour day every working day. Table 9-1 shows the replacement tasks necessary for a typical residential block.

10
Assessment of Project Manpower

For a large multi-million dollar project which spans many years, such as the gas distribution system renewal project of Northwestern Utilities Limited of Alberta, organizing a separate dedicated team to accomplish its completion may be justified. Although some companies assign this task as a part of the responsibilities of the distribution engineer, who may successfully bear this additional responsibility, it really is a matter of evaluating scientifically when and what additional man-hours are needed and whether those man-hours can be performed by the existing staff. If the estimated man-hours are few, shuffling of staff duties may be all that is needed. Defining the project team is an iterative process that has the objective of providing a core project team while making efficient use of existing in-house resources.

The organization chart, Figure 10-1, explains the interrelation of the project team with the gas utility's other existing departments (organizational matrix). Under the task descriptions, proposed work assignments for the staff are delineated. Actual staff requirements are worked out from the planning and scheduling sheet, which is used to first develop the man-months estimate for each category of staff.

METHODOLOGY

The suggested organization (Figure 10-1) is based on the premise that a full-time dedicated team, composed of experienced people, is essential to execute the project. This type of project is important because of its magnitude, its location (usually high-density areas), and its impact on system operation philosophy. In suggesting this organization, it is assumed that the project team will carry out the following work annually:

- Design and construction of 2,000 services, 20 km of various size distribution mains, and 2,000 meter sets to be moved from inside to outside and each fitted with a service regulator.

In order to assess the manpower requirements, a design and construction schedule should first be prepared, in which the project is broken down into tasks. From the tasks, an estimate of man-months by trade should be done. A tally can then be done for each classification (trade), and graphs produced giving man-power loading on a month-to-month basis. (See Figures 10-2 and 10-3.) To account for unproductive time, i.e., vacations and sick time, the total time should be adjusted by 16% or a similarly appropriate number in accordance with your company's known accounting methods.

Figure 10-1. Organizational flow chart.

From the manpower loading graphs, Figure 10-2, an assessment of staff requirements can be done. An organization chart can be developed from the manpower loading graphs. (See Figure 10-1.) For each position, a task de-

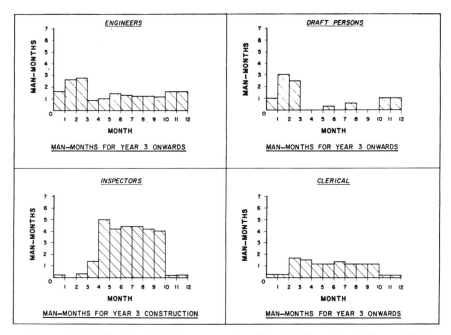

Figure 10-2. Graphs of staff man-months. (For details, refer to Figure 10-3.)

scription giving the duties and qualification requirements should be specified. A search for suitable candidates to fill the positions, either from within
the company or from outside, may follow.

There is another method to assess the manpower requirements that is not
as detailed. That method estimates the overall construction cost first. Then
a suitable percentage is applied to the construction cost to determine the total capital requirements for engineering, drafting, etc. The project engineer
endeavors to keep the cost to within the budgeted figures. (Refer to Cost
Summary in Chapter 7.)

The percentage allowed for engineering design and drafting may vary
from project to project, depending on the industry, the size, and the com-
plexity of work. For a gas distribution system, the engineering design cost
may vary from 5 to 12% of the construction cost. Having determined the
allotted engineering, drafting, and inspection costs, suitable manpower may
be hired as needed based on the project schedule. However, the method
given in this chapter is more accurate and it provides better justification for
management approval.

(text continued on page 102)

No.	DESCRIPTION	YEAR I (Fall Start)

No.	DESCRIPTION	Schedule
I	Project Management - Staff Recruitment, Training, Assignment and Supervision	E .5 .5 .5 .5 .5 .5 .5 Sc .1 .1 .1 .1 .1 .1 .1
2	Develop Procedures on Dead Insertion Technique – Prepare Typical Drawings	D .1 .1 .1 .1 .1 E .5 .5 .5 .5 .5
3	Develop Equipment and Procedures on Live Insertion Technique	E .2 .2 .2 .2 .2
4	Decide New Operating Pressure for Mains	E .1 .1
5	Accurate Quantification of Mains and Services	E .1 Sc .2
6	Pilot Project – Insertion	E .2
7	Update Cost Estimates – Budget Preparation for Following Year	E .1 .2 Sc .1 .1
8	Develop Criteria for Main, Service and Meter Replacement	E .2 .2
9	Develop Criteria for Direct Burial, Dead Insertion, Live Insertion, Replacement of Meters, Treatment of Secondary Services, etc	E .1 .1
10	Prepare Overlay Maps Showing Regulating Stations, LP, MP, IP, and HP Mains. Include Regulating Boxes on System Maps	D .1 .1 E .1 .1
11	Prioritize and Delineate Areas of Importance for Replacement – Review Municipality's Construction Program	E .1 I .1
12	Establish Sequence of Spread of New Higher Pressure System for LP, MP, and IP Lines	D .1 E .1
13	Street Surface Survey for Accurate Cost Information etc	I .1
14	Prepare List of and Contact Large Commercial/Industrial Customers for Future Loads and Meter/Regulating RV Work. Obtain Preliminary Dates for Shutdown.	E .1 .1 .1 I .1 .1 .1
15	Obtain Alignment of Foreign Utilities Near Gas Mains	D .5 E .1 I .1
16	Review Alignments, Decide Suitability of Location for Insertion versus Trenching	E .1
17	Prepare Load Information for Adequate Sizing. Check With Marketing for Future Loads / Developments	E .1
18	Network Analysis: Determine Size and Location of Pipe for Replacement. Identify Redundant Mains. Network Runs to Check Adequacy of Pressure in Surrounding Mains After Successive Step of System Conversion.	E .1 .1 .1

Figure 10-3. Planning and scheduling chart.

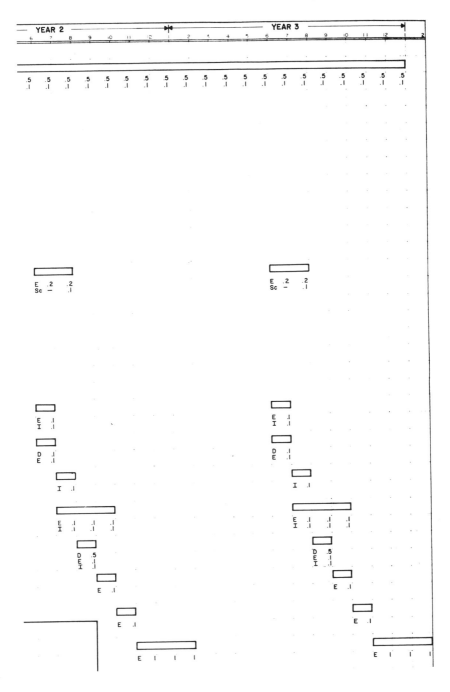

Figure 10-3. Continued.

No.	DESCRIPTION	YEAR I (Fall Start)
19	Prepare Drawings for Regulating Stations Modifications. Raise Work Orders, Order Station Equipment and Material.	
20	Prepare Construction Maps (Tie-ins, Cap Off, Abandonment, Temporary, Supply, etc) Raise Blanket Work Order	
21	Prepare / Update Contract Document for Construction	
22	Invite Bids from Contractors - Select Contractor for Mains, Services, Meter Sets	
23	List Material Needs and Place Requisitions for Pipe, Fittings, Regulators, Meters, Fusion Equipment, CNG Bottles or Trailers.	
24	Construct (Retire or Upgrade) Regulating Stations	
25	Prepare Listing of Affected Customers and Prepare Notices for Customers and the Public, Warning Them of Gas and Traffic Disruptions. Confirm Shutdowns if Appropriate	
26	Award of Contract for Mains / Services / Meter Sets Relocation	
27	Construct Mains, Services, Meter Sets. Shutdowns and Sequence of Replacement as per Plan	
28	As Built Plotting	

|← YEAR I →|←

DRAFT PERSON	- D	0.1	1.2	1.6	—	0.1	—	—
ENGINEER	- E	1.2	2.2	1.9	1.7	2.2	1.5	1.5
INSPECTOR	- I	—	0.3	0.2	0.3	—	—	—
SECRETARIAL / CLERK	- Sc	2.1	0.1	0.2	0.2	0.1	0.1	0.1
TOTAL			3.4	3.8	3.9	2.2	2.4	1.6	1.6

	STUDY — YEAR Nº I	PARTIAL CONSTRUCTION — YEAR 2	FULL PRODUCTION — YEAR 3	STAFF NEED FOR YEAR 3 ONWARDS	
D	2.9 + 16% = 3.36 MM	6.7 + 16% = 7.8 MM	9.1 + 16% = 10.6 MM	D = 10.6 MM	SAY I DRAFT PERSON
E	5.3 + 16% = 6.15 MM	20.5 + 16% = 23.8 MM	17.8 + 16% = 20.6 MM	E = 20.6 MM	SAY 2 ENGINEERS
I	0.5 + 16% = 0.6 MM	14.3 + 16% = 16.6 MM	27.4 + 16% = 31.8 MM	I = 31.8 MM	SAY 3 INSPECTORS
Sc	2.4 + 16% = 2.8 MM	5.6 + 16% = 6.5 MM	10.4 + 16% = 12.0 MM	Sc = 12.0 MM	SAY I SECRETARY / CLERK
TOTAL MAN MONTHS	12.9 MM	54.7 MM	75.0 MM	7 PERSONS	

Figure 10-3. Continued.

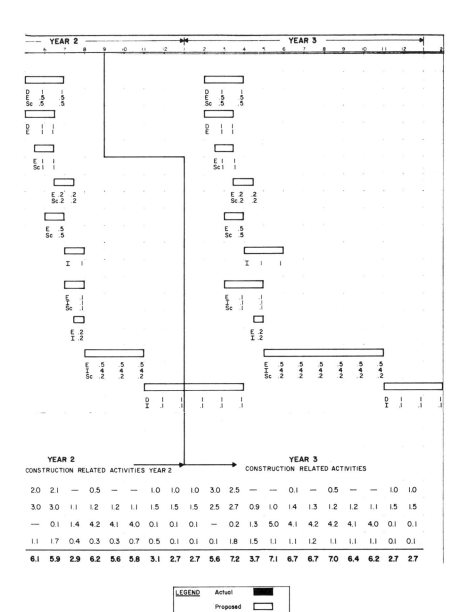

Figure 10-3. Continued.

(text continued from page 97)

ORGANIZATIONAL CONCEPT

In evolving the organization chart, four things must be kept in view:

1. Management's policy, if any, on promotions from within and its over-all game plan on organizational structure.

2. Positions of a similar nature should be grouped together for efficient operation.

3. The company's existing resources should be used wherever possible.

4. Coordination of field functions, engineering, and other company departments with contractors and other utilities/governmental bodies, is essential.

Keeping these in mind, two sub-groups are proposed, a design group and a construction group. Both groups should report to the project engineer for administrative purposes. For functional responsibilities, i.e., day-to-day production, the design group will seek direction from the project engineer.

To inspect the work of a crew of as many as 25 people, a full-time on-site inspector is required. At least two construction crews will work simultaneously on this project. This number may go up to four, depending on the construction activity. For the inspection of the secondary lines and meter replacement work, a customer-service inspector is required.

All construction-related activities are seasonal. Construction is expected to commence in May and end in October every year. The requirement for winter shutdown is necessary in Alberta, as the ground freezes in winter, making it very expensive and time-consuming to carry out outdoor construction work. For areas such as Southern United States where construction work may continue throughout the year, scheduling of work and manpower needs must be adjusted accordingly. The permanent positions of project engineer, secretary, and design engineer will work steadily with some seasonal variations, but no slack time is envisaged.

Positions described in the task descriptions are marked as permanent (P) or seasonal (S). Adjustments should be made for each project team, as the positions and task descriptions given here are illustrations only. Most companies have an existing organizational structure and policies to which they must conform.

TASK DESCRIPTIONS

The task descriptions given will serve as examples. Positions and man-hours are based on a distribution renewal project for 200 km of mains and 18,000 services, to be completed over 10 years. Refer to Chapter 2 for further discussion.

Project Engineer (P)

Special needs: Car with company radio.
Supervisor: Manager of Engineering.
Qualifications:
    Education: Registration as P.E.
    Experience: 10 years gas-utility experience, 5 of which must have been spent in a supervisory role in the design, construction, or operation of gas systems, directing the work of other technical staff.

Duties:

1. Provide technical and management guidance to the distribution system renewal team.
2. Motivate staff to meet company's budgetary and construction targets. Maintain high morale among team members.
3. Maintain good intercompany communications. Coordinate the efforts of team members and other respective groups/departments.
4. Carry responsibility for overall performance and timely completion.
5. Prepare the budget and maintain cost control.
6. Select and prioritize areas for replacement program.
7. Carry out administrative duties, i.e., progress reports, organization, staff selection and assignment, development, and deployment.
8. Develop technical standards and procedures applicable to this project.
9. Carry out research and development necessary for live- and dead-insertion techniques.
10. Prepare and amend contract documents, when necessary, for work relating to this project. Invite and assess bids, recommend contractors, and review performance. Verify contractor's invoices for payment.

Design Engineer (P)

Special needs: None.
Supervisor: Project Engineer.
Qualifications:
    Education: Registration as P.E.
    Experience: 5 years in gas industry, of which at least 1 year must be in
           network analysis and 1 year in regulation.
Duties:

1. Obtain and update customer load information.
2. Obtain overlay maps showing regulating stations, LP, MP, IP, and HP mains.
3. Establish sequence of spread of new higher-pressure system.
4. Prepare list of and contact large commercial/industrial customers for future loads.
5. Review alignment of foreign utilities near gas mains and decide on the method of replacement, i.e., insertion or direct burial.
6. Carry out network analysis to size the new mains system.
7. Prepare construction maps.
8. Prepare bills of materials.
9. Prepare estimate and raise work orders.
10. Determine the need and arrange for temporary gas supply to the customers, i.e., CNG bottles, trailers, etc.
11. Modify and design new regulating stations.
12. Miscellaneous tasks as required by the supervisor.

Secretary/Clerk (P)

Special needs: None.
Supervisor: Project Engineer.
Qualifications:
    Education: High school diploma, word processing course.
    Experience: One year experience using a word processor. Familiarity
           with piping material.

Duties:

1. Typing, using an electric typewriter and/or word processing equipment.
2. Receive and make phone calls, take messages, and prepare mailings.

3. Filing, photocopying, etc.
4. Prepare tallies of mains and services.
5. Prepare requisitions and place orders.
6. Expedite deliveries.
7. Scrutinize invoices for payment to vendors and contractors.
8. Miscellaneous tasks as required by the supervisor.

Inspector/As-Built Person (S)

Special needs: 1/2-ton truck with company radio.
Supervisor: Field Supervisor-Contract Inspections. Reports to Project Engineer for project-related items, but reports to Field Supervisor-Contract Inspections for administrative purposes.
Qualifications:
  Education: High school diploma, PE pipe fusion ticket, 3 months drafting training or equivalent.
  Experience: 5 years gas-mains construction experience, 2 of which must be as an inspector.
Duties: The person in this position is responsible for the quality control of all mains and service construction work up to meter stopcock.

1. Act as company's representative in the field.
2. Maintain compliance with codes and the company's policies.
3. Ensure good workmanship and control the quality of the workmanship and the materials to be used.
4. Maintain a daily log of contractor's activities and the progress of the work.
5. Provide field interpretation of construction-related problems such as alignment, cover, code requirements, and safety matters.
6. Prepare field as-built drawings of the work completed.
7. Verify contractor's work for payment.
8. Miscellaneous tasks as required by the supervisor.

Inspector-Customer Service (S)

Special needs: 1/2-ton truck with company radio.
Supervisor: Field Supervisor-Contract Inspections. Reports to Project Engineer for project-related items, but reports to Field Supervisor-Contract Inspections for administrative purposes.
Qualifications:
  Education: High school diploma, Gas fitter's license.

Experience: 3 years as a licensed gas fitter, 2 years as an inspector.
Duties: The person in this position is responsible for the quality control of
all construction work downstream of the meter stopcock.

1. Liaise shutdown and relighting work with customers and contractors.
2. Inspect and ensure compliance of contractor's work to company's requirements.
3. Witness tests and tag equipment, if necessary.

ESTIMATE OF MAN-MONTHS

A man-months estimate is based on task breakdown as given in the planning and scheduling chart (Figure 10-3). For ready reference, information on man-months for year 3 onwards is repeated below.

Engineer	20.6 man-months = 2 man-years
Draft person	10.6 man-months = 1 man-year
Inspector	31.8 man-months = 3 man-years
Secretarial-clerical	12 man-months = 1 man-year

11
Work Measurement and Contractor Payment

A cost analysis for contract work and work done by a company's construction department should be carried out to determine the most cost-effective way of renewing the distribution system piping. When comparing contract costs with in-house construction costs, hidden company charges (such as heavy equipment, fleet charges, payroll benefits, and other company overheads) should also be taken into account.

Typically, the gas utility's equipment and overhead charges for pipe replacement work are as follows:

Construction equipment and fleet charge	= 30 to 40% of direct labor cost
Company overheads	= 50 to 60% of direct labor cost
Total other costs in addition to direct labor	= 80 to 100% of direct labor charges

Gas utilities usually provide their own materials, engineering design, and project management. Hence, costs attributed to these items remain the same whether the construction work is done by the contractor or by the company's construction crew. Besides considering the hidden overhead costs as described above, if the work is of a seasonal nature, as is the case for the northern U.S. and Canada where outdoor construction activities are slowed or completely halted for the winter, labor carryover costs during this period of shutdown must also be considered. If it is to the company's advantage to contract the work, the tables of unit rates and day work should be completed.

ASSUMPTIONS

When developing this recommended method of payment, it is assumed that:

1. The renewal project is of sufficient size that contracting out the work is beneficial to the company.
2. Urban residential, commercial, and industrial areas are involved and, due to the variations in the street surfaces, depths of cover, sizes, materials, etc., bidding on a lump-sum basis is not practical. Thus, the contractor will be paid as per a schedule of unit rates.
3. Pipe replacement work will be done using the dead insertion and direct burial methods. The schedule of unit rates for live insertion is not covered in this book.

DEVELOPMENT OF UNIT RATES SCHEDULE

The project should be broken down into tasks and sub-tasks. Before this can be done, it is necessary to develop procedures and finalize the scope of work and pertinent quantities of work items. Typically, in a gas distribution renewal project, contract work may be classified under the following categories:

- Mains
- Services
- Customer-service related, i.e., meter set and piping downstream of the meter set

Public relations and government matters are usually dealt with by the company itself and not by the contractor. Each of the above categories may then further be broken down as follows:

Mains

1. Steel pipe—(new)
2. Plastic pipe by trenching—(new)
3. Plastic pipe by insertion—(new)
4. Miscellaneous mains and service-related pay items
5. Upgrading existing steel main
6. Upgrading existing plastic main

Services

1. Steel pipe—(new)
2. Plastic pipe by trenching, boring, plowing, etc.—(new)
3. Plastic pipe by insertion—(new)
4. Upgrading existing steel services
5. Upgrading existing plastic services

Customer-Service Related

1. Meter and regulator removal/reinstallation
2. Internal piping alternation
3. Secondary services (new or alteration)

Schedule of Unit Rates

One gas utility developed the pay items listed in Table 11-1 to cover the entire scope of its construction work using contractors to do mains, services, and customer-service work, and using the direct burial as well as the insertion technique. You will notice that the schedule of unit rates also contains certain pay items to accommodate the use of plastic sleeves and plastic Thinsulators for the insertion of plastic pipe into existing metal pipe.

This list suits urban gas-distribution system renewal work specifically. Some pay items refer to "figures." These figures are in Chapter 13.

(text continued on page 120)

Table 11-1
Schedule of Unit Rates

Item No.	Description	Size mm	Estimated Quantity	Unit Price $	Total Price $
Main Installation					
1.2	Plastic main by trenching				
	For the complete	26.7	_____	_____/m	_____
	installation of the plastic	42.2	_____	_____/m	_____
	underground mains as	60.3	_____	_____/m	_____
	detailed in the figures.	88.9	_____	_____/m	_____
		114.3	_____	_____/m	_____
		168.3	_____	_____/m	_____

(Table 11-1 continued on next page)

**Table 11-1 Continued
Schedule of Unit Rates**

Item No.	Description	Size mm	Estimated Quantity	Unit Price $	Total Price $
1.3	Steel main by trenching				
	For the complete	26.7	_____	_____/m	_____
	installation of the welded	42.2	_____	_____/m	_____
	steel underground mains	60.3	_____	_____/m	_____
	as detailed in the figures.	88.9	_____	_____/m	_____
		114.3	_____	_____/m	_____
		168.3	_____	_____/m	_____
		219.1	_____	_____/m	_____
		273.1	_____	_____/m	_____
2.0	Plastic main by dead insertion				
2.1	For the complete	26.7	_____	_____/m	_____
	installation of the fused	42.2	_____	_____/m	_____
	PE underground mains*	60.3	_____	_____/m	_____
		88.9	_____	_____/m	_____
		114.3	_____	_____/m	_____
		168.3	_____	_____/m	_____
		219.1	_____	_____/m	_____
2.2	Install thinsulators as	26.7	_____	_____/ea	_____
	specified in figures.	42.2	_____	_____/ea	_____
		60.3	_____	_____/ea	_____
		88.9	_____	_____/ea	_____
		114.3	_____	_____/ea	_____
		168.3	_____	_____/ea	_____
2.3	Install pipe anchor.	42.2	_____	_____/ea	_____
	Rate based on PE carrier	60.3	_____	_____/ea	_____
	pipe size.	88.3	_____	_____/ea	_____
		114.3	_____	_____/ea	_____
		168.3	_____	_____/ea	_____
		219.1	_____	_____/ea	_____
2.4	Install plastic sleeve.	42.2	_____	_____/m	_____
	Rate based on PE carrier	60.3	_____	_____/m	_____
	pipe size.	88.9	_____	_____/m	_____
		114.3	_____	_____/m	_____
		168.3	_____	_____/m	_____
		219.1	_____	_____/m	_____
2.5	Install steel main valve	42.2	_____	_____/ea	_____
	and valve box in PE	60.3	_____	_____/ea	_____
	main.	88.9	_____	_____/ea	_____
		114.3	_____	_____/ea	_____
		168.3	_____	_____/ea	_____

Table 11-1 Continued
Schedule of Unit Rates

Item No.	Description	Size mm	Estimated Quantity	Unit Price $	Total Price $
2.6	Install PE main valve and valve box.	42.2	_____	_____/ea	_____
		60.3	_____	_____/ea	_____
		88.9	_____	_____/ea	_____
		114.3	_____	_____/ea	_____
2.7	Install steel valve and valve box in steel main.	168.3	_____	_____/ea	_____
		219.1	_____	_____/ea	_____
		273.1	_____	_____/ea	_____
3.0	Miscellaneous M&S-related				
3.1	Concrete removal—slabs				
	To cut and remove concrete roadways, sidewalks, parking lots, etc.				
	Up to 150 mm thickness		_____m	_____/m	_____
	151 mm to 225 mm		_____m	_____/m	_____
	226 mm to 300 mm		_____m	_____/m	_____
3.2	Concrete removal— reinforced slabs				
	To cut and remove concrete roadways, sidewalks, parking lots, etc.				
	Up to 150 mm thickness		_____m	_____/m	_____
	151 mm to 225 mm		_____m	_____/m	_____
	226 mm to 300 mm		_____m	_____/m	_____
3.3	Concrete removal—curb or gutters				
	To remove concrete curbs, and/or gutters, lineal length.		_____m	_____/m	_____
3.4	Interlocking brick (stone) removal				
	To remove interlocking brick (stone) sidewalks, driveways, or streets.		_____m²	_____/m²	_____
3.9	Frost depth				
	0.3 m to 0.6 m		_____m	_____/m	_____
	0.61 m to 1 m		_____m	_____/m	_____

Table 11-1 Continued
Schedule of Unit Rates

Item No.	Description	Size mm	Estimated Quantity	Unit Price $	Total Price $
3.10	Crushed gravel				
	Supply and place crushed gravel, in 150 mm layers.		_____m²	_____/m²	_____
3.11	Padding material (sand)				
	Supply and place padding material, 150 mm above and below pipe.	42.2–114.3 168.3–323.9	_____m _____m	_____/m _____/m	_____ _____
3.12	Sod/seeding				
	To cut, take up, and replace sod or supply and place new sod or seeding.		_____m² _____m²	_____/m² _____/m²	_____ _____
3.13	Asphalt removal				
	To cut and remove asphalt roadway, sidewalks, parking lots, etc., Up to 150 mm thickness. 151 mm to 225 mm 226 mm to 300 mm		_____m _____m _____m	_____/m _____/m _____/m	_____ _____ _____
3.15	Concrete replacement—slab				
	Supply and place concrete in roadway, sidewalk, parking lot, driveway. Thickness	Up to 100 mm 101–150 mm 151–225 mm	_____m² _____m² _____m²	_____/m² _____/m² _____/m²	_____ _____ _____
3.16	Concrete replacement—reinforced slab				
	Supply and place reinforced concrete in roadway, sidewalk, parking lot, driveway. Thickness	Up to 100 mm 101–150 mm 151–225 mm	_____m² _____m² _____m²	_____/m² _____/m² _____/m²	_____ _____ _____

Table 11-1 Continued
Schedule of Unit Rates

Item No.	Description	Size mm	Estimated Quantity	Unit Price $	Total Price $
3.17	Concrete replacement—curb and/or gutter				
	Supply and place		_____ m	_____ /m	_____
3.18	Interlocking brick replacement				
	Supply and place		_____ m²	_____ /m²	_____
3.19	Asphalt replacement				
	Supply and place asphalt, roadways, laneways, parking lots, driveways.				
	Up to 100 mm thickness		_____ m²	_____ /m²	_____
	101–150 mm thickness		_____ m²	_____ /m²	_____
	151–225 mm thickness		_____ m²	_____ /m²	_____
3.20	Cold mix placement				
	Supply and place cold mix on sidewalks, laneways, parking lots, etc.				
	100 mm thick		_____ m²	_____ /m²	_____
3.21	Cased crossing Highway, waterway, and railway crossings				
	For the installation of carrier pipe, casing, casing seal bushings, casing insulators, vents, warning signs, painted posts. Carrier sizes as given.	42.2	_____ m	_____ /m	_____
		60.3	_____ m	_____ /m	_____
		88.9	_____ m	_____ /m	_____
		114.3	_____ m	_____ /m	_____
		168.3	_____ m	_____ /m	_____
		219.1	_____ m	_____ /m	_____
		273.1	_____ m	_____ /m	_____
		323.9	_____ m	_____ /m	_____
3.22	Uncased crossing				
	Boring or driving for mains (without casing) under concrete pad, driveway, roadways, railways, or watercourse, and laying carrier pipe.	26.7	_____ m	_____ /m	_____
		42.2	_____ m	_____ /m	_____
		60.3	_____ m	_____ /m	_____
		88.9	_____ m	_____ /m	_____
		114.3	_____ m	_____ /m	_____
		168.3	_____ m	_____ /m	_____
		219.1	_____ m	_____ /m	_____
		273.1	_____ m	_____ /m	_____

Table 11-1 Continued
Schedule of Unit Rates

Item No.	Description	Size mm	Estimated Quantity	Unit Price $	Total Price $
3.24	Foreign crossing—direct burial only				
	For crossing cable, pipe, or any other foreign structure (max 1.5 m deep) where excavating by hand is required.		_____	_____/ea	_____
3.29	Abnormal extra-depth ditch—for direct burial only				
	For excavating abnormal extra-depth ditch other than normal-depth ditch (over 1.5 m deep and 5 m in length).				
	Extra depth	0–150 mm	_____m	_____/m	_____
		151–300 mm	_____m	_____/m	_____
		301–450 mm	_____m	_____/m	_____
		451–600 mm	_____m	_____/m	_____
		601–750 mm	_____m	_____/m	_____
		751–900 mm	_____m	_____/m	_____
		901–1,200 mm	_____m	_____/m	_____
3.30	Installation of a cathodic protection test point and anode.		_____	_____/ea	_____
3.31	Anode only installation on main, service, or tracer wire and associated cadwelding.		_____	_____/ea	_____
3.35	Installation of stopper fitting on steel pipe, for MP and IP steel line, weld on fitting.	42.2	_____	_____/ea	_____
		60.3	_____	_____/ea	_____
		88.9	_____	_____/ea	_____
		114.3	_____	_____/ea	_____
		168.3	_____	_____/ea	_____
		219.1	_____	_____/ea	_____
		273.1	_____	_____/ea	_____
		323.9	_____	_____/ea	_____
3.36	Drilling MP and IP steel lines and stopping gas flow.	42.2	_____	_____/ea	_____
		60.3	_____	_____/ea	_____
		88.9	_____	_____/ea	_____
		114.3	_____	_____/ea	_____
		168.3	_____	_____/ea	_____

Table 11-1 Continued
Schedule of Unit Rates

Item No.	Description	Size mm	Estimated Quantity	Unit Price $	Total Price $
		219.1	_____	_____/ea	_____
		273.1	_____	_____/ea	_____
		323.9	_____	_____/ea	_____
3.37	Installing steel bypass between line stopper fittings.	42.2	_____	_____/ea	_____
		60.3	_____	_____/ea	_____
		88.9	_____	_____/ea	_____
		114.3	_____	_____/ea	_____
3.38	Remove steel main from trench, cut and transport to company's pipe yard.	88.9	_____m	_____/m	_____
		114.3	_____m	_____/m	_____
		168.3	_____m	_____/m	_____
		219.1	_____m	_____/m	_____
		273.1	_____m	_____/m	_____
		323.9	_____m	_____/m	_____
4.0	Upgrading existing mains. (Service lines included.)				
4.1	Excavate existing steel main, disconnect ends, pressure test, reconnect, and backfill.	42.2	_____test	_____/test	_____
		48.3	_____test	_____/test	_____
		60.3	_____test	_____/test	_____
		88.9	_____test	_____/test	_____
		114.3	_____test	_____/test	_____
		168.3	_____test	_____/test	_____
		219.1	_____test	_____/test	_____
		273.1	_____test	_____/test	_____
		323.9	_____test	_____/test	_____
		406.4	_____test	_____/test	_____
4.2	Excavate existing plastic main, disconnect ends, pressure test, reconnect, and backfill.	42.2	_____test	_____/test	_____
		48.3	_____test	_____/test	_____
		60.3	_____test	_____/test	_____
		88.9	_____test	_____/test	_____
		114.3	_____test	_____/test	_____

Subtotal for all main installation items, 1.1 to 4.2.   _____

Service Installation

5.1   Plastic service by methods other than insertion, i.e., trenching, boring, plowing, hand digging, etc., or any combination thereof

Table 11-1 Continued
Schedule of Unit Rates

Item No.	Description	Size mm	Estimated Quantity	Unit Price $	Total Price $
	Up to 25 linear meters long				
	To install PE service	15.9	_____	_____/ea	_____
	line from meter-stop	26.7	_____	_____/ea	_____
	or template to the	42.2	_____	_____/ea	_____
	main. Excavate,	60.3	_____	_____/ea	_____
	place, test, purge	88.9	_____	_____/ea	_____
	and backfill.	114.3	_____	_____/ea	_____
	Over 25 linear meters long				
		15.9	_____m	_____/m	_____
		26.7	_____m	_____/m	_____
		42.2	_____m	_____/m	_____
		60.3	_____m	_____/m	_____
		88.9	_____m	_____/m	_____
		114.3	_____m	_____/m	_____
5.2	Plastic service by dead insertion**				
	Up to 25 linear meters	15.9	_____	_____/ea	_____
	long. To install PE	26.7	_____	_____/ea	_____
	service line from meter-	42.2	_____	_____/ea	_____
	stop or template to the	60.3	_____	_____/ea	_____
	main by insertion				
	method.	88.9	_____	_____/ea	_____
5.3	Plastic service by dead insertion**				
	Over 25 linear meters	15.9	_____m	_____/m	_____
	long. To install PE	26.7	_____m	_____/m	_____
	service line as above.	42.2	_____m	_____/m	_____
		60.3	_____m	_____/m	_____
		88.9	_____m	_____/m	_____
5.4	Steel service by trenching				
	Up to 25 linear	15.9	_____	_____/ea	_____
	meters long. To	26.7	_____	_____/ea	_____
	install steel service	42.2	_____	_____/ea	_____
	line from meter-stop	60.3	_____	_____/ea	_____
	or template to the	88.9	_____	_____/ea	_____
	main. (Excavate,	114.3	_____	_____/ea	_____
	place, test, purge, and backfill.)				

Table 11-1 Continued
Schedule of Unit Rates

Item No.	Description	Size mm	Estimated Quantity	Unit Price $	Total Price $
5.5	Steel service by trenching				
	Over 25 linear meters long. To install steel service line as above.	15.9	_____	_____/m	_____
		26.7	_____	_____/m	
		42.2	_____	_____/m	_____
		60.3	_____	_____/m	_____
		88.9	_____	_____/m	_____
5.6	Curb valve				
	Install curb valve and box. Stencil location of box on fence/garage.	15.9	_____	_____/ea	_____
		26.7	_____	_____/ea	_____
		42.2	_____	_____/ea	_____
		60.3	_____	_____/ea	_____
		88.9	_____	_____/ea	_____
		114.3	_____	_____/ea	_____
5.7	Outside riser				
	Install aboveground outside riser.	15.9	_____	_____/ea	_____
		26.7	_____	_____/ea	_____
		42.2	_____	_____/ea	_____
		60.3	_____	_____/ea	_____
		88.9	_____	_____/ea	_____
		114.3	_____	_____/ea	_____
5.8	Riser protector				
	Fabricate and install riser protector (material company supplied).		_____	_____/ea	_____
5.9	Disconnect service				
	Excavate, disconnect, backfill.	26.7	_____	_____/ea	_____
		42.2	_____	_____/ea	_____
		60.3	_____	_____/ea	_____
		88.9	_____	_____/ea	_____
6.0	Upgrading existing services. (Main not upgraded.)				
6.1	Steel service—excavate, bell hole, disconnect, pressure test, reconnect, backfill.	15.9	_____	_____/ea	_____
		26.7	_____	_____/ea	_____
		42.2	_____	_____/ea	_____
		60.3	_____	_____/ea	_____
		88.9	_____	_____/ea	_____

Table 11-1 Continued
Schedule of Unit Rates

Item No.	Description	Size mm	Estimated Quantity	Unit Price $	Total Price $
6.2	Plastic service—	15.9	_____	_____/ea	_____
	excavate bell hole,	26.7	_____	_____/ea	_____
	disconnect, pressure	42.2	_____	_____/ea	_____
	test, reconnect,	60.3	_____	_____/ea	_____
	backfill.	88.9	_____	_____/ea	_____
	Subtotal for all service installation items, 5.1 to 6.2				_____

Customer-Service Related

Item No.	Description	Size mm	Estimated Quantity	Unit Price $	Total Price $
7.0	Downstream from meter-stop				
7.1	Remove meter set	33.4	_____	_____/ea	_____
	at building. Rate	42.2	_____	_____/ea	_____
	based on meter	48.3	_____	_____/ea	_____
	inlet size.	60.3	_____	_____/ea	_____
		88.9	_____	_____/ea	_____
		114.3	_____	_____/ea	_____
7.2	Core through wall and plug off redundant hole with mortar where required.		_____	_____/ea	_____
7.3	Hang meter set.	33.4	_____	_____/ea	_____
	Rate based on inlet	42.2	_____	_____/ea	_____
	size. Up to 2 m of	48.3	_____	_____/ea	_____
	piping on either side	60.3	_____	_____/ea	_____
	of meter set, from	88.9	_____	_____/ea	_____
	riser to customer's	114.3	_____	_____/ea	_____
	piping through wall.				
7.4	Multi-meter header installation, 26.7 mm or 33.4 mm.				
	2-meter header		_____	____install	_____
	3-meter header		_____	____install	_____
	4-meter header		_____	____install	_____
7.5	Fabricate and assemble	33.4	_____m	_____/m	_____
	internal screwed piping	42.2	_____m	_____/m	_____
	and provide pipe	48.3	_____m	_____/m	_____
	support as needed.	60.3	_____m	_____/m	_____
	Valve, fitting, and	88.9	_____m	_____/m	_____
	testing included.				

Table 11-1 Continued
Schedule of Unit Rates

Item No.	Description	Size mm	Estimated Quantity	Unit Price $	Total Price $
7.6	Fabricate and assemble internal welded piping and provide pipe support as needed. Valve, fitting, and testing included.	33.4 42.2 48.3 60.3 88.9 114.3	_____m _____m _____m _____m _____m _____m	_____/m _____/m _____/m _____/m _____/m _____/m	_____ _____ _____ _____ _____ _____
7.7	Test and reconnect inside piping.		_____	_____/test	_____
7.8	Relight customer's appliances. Do safety check.		_____	_____/prem	_____
7.9	Underground secondary service piping. Excavate, lay, connect, test, and backfill secondary service pipe.	26.7 42.2	_____m _____m	_____/m _____/m	_____ _____
7.10	Fabricate and install riser return bend and install link seal in cored wall.	26.7 42.2 60.3 88.9	_____ _____ _____ _____	_____/ea _____/ea _____/ea _____/ea	_____ _____ _____ _____
	Subtotal for all customer-service related items, 7.1 to 7.10.				_____

* Transport pipe and fittings from company to job site and store them. Supply labor and equipment for the excavations required to perform plastic-main insertion. Install and weld all fittings needed on existing steel line for gas shut-off, etc. Fuse plastic pipe and fittings. Cut necessary length of existing steel pipe. Clean steel main internally. Insert plastic pipe and tracer wire. Pressure test. Remove surplus excavated material off job site. Tie-in and purge plastic main. Backfill holes, clean up, and return all surplus material to company's stores.

** Transport pipe and fittings from company to job site and store them. Supply labor and equipment for the excavations required to perform plastic service insertion. Cut out a section of main containing service tee. Cut out old service pipe from main up to and including curb valve. Remove outside old piping upstream of the meter set at building. Clean old steel line. Insert plastic line and tracer wire. Install riser and shutoff valve at building. Tie-in at main. Fuse service tee, test service, tap main and purge service line. Backfill at building. Clean up, replace plants and sod on private property. Part trenching and part insertion shall be paid at the appropriate insertion rate.

(text continued from page 109)

FORCE ACCOUNT

The force account sets hourly/daily rates for the contractor's manpower and equipment. (See Table 11-2.) Since there is no incentive for the contractor to finish the work early when he is paid by the force-account method, it should only be used if the work is not covered by a pay item or because a suitable lump sum or unit price cannot be negotiated. The force account should be viewed as the last resort as a method of work measurement and payment. However, it is an essential component of a contract document, as it covers unforeseen situations and undefinable tasks. Force-account rates should be obtained from the contractor at the same time as other unit rates are obtained.

Table 11-2
Force Account Equipment Rates for Extra Work

The following list shows the equipment which the contractor will furnish. The rates indicated cover all costs, including taxes, insurance, fuel, maintenance and repairs, overhead, and profit. The contractor shall indicate size and type of equipment for rates and shall not include operator's rate.

Equipment	Rate Per Hour	Rate Per 10-Hour Day
1. Trenching machine	$_____	$_____
2. Trenching machine	$_____	$_____
3. Trenching machine	$_____	$_____
4. Backhoe $3/4$ m^3	$_____	$_____
5. Backhoe $1/2$ m^3	$_____	$_____
6. Backhoe hydraulic 1 m^3	$_____	$_____
7. Backhoe	$_____	$_____
8. Lowboy, hiboy (truck & trailer _____t)	$_____	$_____
9. Truck with pipe trailer _____t	$_____	$_____
10. Truck with pipe trailer _____t	$_____	$_____
11. Truck, winch _____t	$_____	$_____
12. Truck, winch	$_____	$_____
13. Truck c/w hydraboom, trailer _____t	$_____	$_____
14. Truck, flatbed	$_____	$_____
15. Truck, fuel	$_____	$_____
16. Truck, grease	$_____	$_____
17. Bus _____ passenger	$_____	$_____
18. Truck, panel	$_____	$_____
19. Truck $1/2$ t	$_____	$_____
20. Truck $1/4$ t	$_____	$_____

Table 11-2 Continued
Force Account Equipment Rates for Extra Work

Equipment	Rate Per Hour	Rate Per 10-Hour Day
21. Fittings truck equipped with small tools, and fusion and testing equipment	$_____	$_____
22. Welding rig, truck, welding machine & misc. equipment	$_____	$_____
23. Air compressor with tools _____kPa _____m³/min.	$_____	$_____
24. Air compressor with tools _____kPa _____m³/min.	$_____	$_____
25. Pumps (Specify type, make, and capacity)	$_____	$_____
26. Barricades		$_____
27. Traffic flasher		$_____
28. Steel plates—up to 1.5 m × 3 m × 25 mm		$_____

Manpower Rate for Extra Work

The hourly rates set out below include taxes, insurance, overhead and profit. These rates are established on the basis that no increase or change in the rates will be made for overtime hours or for any other reason.

Classification	Hourly Rates
1. Dozer operator	$_____
2. Backhoe operator	$_____
3. Compressor operator (up to 10.3 m³/min.)	$_____
4. Compressor operator (10.3 m³/min. and over)	$_____
5. Welder	$_____
6. Welder's helper	$_____
7. Fusion operator	$_____
8. Swamper	$_____
9. Jackhammer operator	$_____
10. Powersaw operator	$_____
11. Truck driver, lowboy, hiboy	$_____
12. Truck driver, winch truck	$_____
13. Truck driver, flatbed	$_____
14. Laborer	$_____
15. Watchman	$_____
16. Foreman	$_____
17. Superintendent	$_____
18. Engineer	$_____
19. Subsistence per day per worker	$_____

Date _____   Bidder _____

By _____

DAILY/WEEKLY WORK MEASUREMENT AND PROGRESS PAYMENT REPORT

The progress payment report form (Figure 11-1) is a multi-function form which can be used to:

1. Itemize progress reports for daily work.
2. Itemize progress reports for weekly summaries.
3. Get the company inspector's approval for work completed.
4. Invoice by contractor and get payment approval by the company.

Processing of Progress Payment Report

The progress payment report form itemizes all the pay items, and should be prepared in quadruplicate (with self-carbon feature) depending upon the needs of a particular company.

The suggested paper flow follows. Also see Figure 11-2.

1. The contractor will complete triplicate forms, filling out the appropriate boxes and the pay item quantities for each day's construction, and will obtain the company inspector's approval and give the original to the company. The contractor shall retain the other two copies. Weekly total column or any price information shall not be completed at this time.
2. On a weekly (or a similarly convenient period) basis, the contractor shall prepare the forms in quadruplicate, this time completely filling out the top portion and the weekly quantity columns only. He will attach a copy of each of the previously signed daily reports already in his possession with the weekly report, and obtain inspector's signatures for the totals.
3. The contractor will leave the original weekly sheet (and the attachments) with the company inspector and pass on the remaining three copies to his office for completion of the last six columns, which pertain to unit prices, total quantities, and the total cost. The office will also fill out the last three rows of the form for the invoice amount.
4. Two copies of the completed form shall be sent directly to the company's construction manager for payment. The company's construction manager or his designate shall compare the invoice copies against the original copy in the inspector's possession, compare the contract document for prices, and approve and pass the invoice to accounting for payment. This method ensures that access to bid prices is limited to authorized persons only.   *(text continued on page 128)*

PROGRESS PAYMENT REPORT

DAILY/WEEKLY

PAGE ___ OF ___
SHEET A: MAINS

CONTRACTOR	CONTRACT No.	PLANT No.	W.O. No.	INVOICE No.	PAY PERIOD
STREET (BLOCK) ADDRESS					DRAWING No.
PREPARED BY / DATE	ACCEPTED BY / DATE	CHECKED BY Co. / DATE	Co. APPROVAL / DATE	Co. APPROVAL / DATE	Co. APPROVAL / DATE

PAY ITEM No.	DESCRIPTION	SIZE mm	DAILY QUANTITIES (FOR DAILY REPORT ONLY) MON	TUES	WED	THURS	FRI	SAT	SUN	TOTAL QTY THIS PERIOD	UNIT	FOR CONTRACTOR'S USE: UNIT PRICE	COST THIS PERIOD	TOTAL QUANTITIES: PREVIOUS	TO DATE	FOR CONTRACTOR'S USE: TOTAL COST: PREVIOUS	TO DATE
1.2	Plastic Main by Trenching										m						
											m						
											m						
1.3	Steel Main by Trenching										m						
											m						
											m						
2.1	Plastic Main by Dead Insertion										m						
											m						
											m						
2.4	Plastic Sleeve										m						
											m						
											m						
3.1	Concrete Removal	150mm									m^2						
											m^2						
3.3	Remove Concrete Curb or Gutter										m^2						
3.12	Remove and Replace Sod, or Seeding										m^2						
3.13	Cut and Remove Asphalt	150mm									m^2						
											m^2						
3.15	Concrete Replacement- Slab										m^2						
											m^2						
3.17	Concrete Replacement- Curb, Gutter										m						
3.19	Asphalt Replacement	100mm									m^2						
											m^2						
3.22	Uncased Crossing Boring or Driving for Main Without Casing										m						
											m						
3.24	Foreign Crossing										ea						
3.25											ea						
3.3	Anode and Test Point										ea						
3.31	Anode										ea						
	TOTAL FOR THIS INVOICE:																
	HOLD BACK:																
	PAY OUT:																

Figure 11-1. Progress payment report.

(Figure 11-1 continued on next 3 pages)

Figure 11-1. Continued

PROGRESS PAYMENT REPORT

DAILY/WEEKLY

PAGE ___ OF ___
SHEET C: SERVICES

CONTRACTOR	CONTRACT No.	PLANT No.	W.O. No.	INVOICE No.	PAY PERIOD
STREET (BLOCK) ADDRESS					DRAWING No.
PREPARED BY DATE	ACCEPTED BY DATE	CHECKED BY Co. DATE	Co. APPROVAL DATE	Co. APPROVAL DATE	Co. APPROVAL DATE

PAY ITEM No.	DESCRIPTION	SIZE mm	DAILY QUANTITIES (FOR DAILY REPORT ONLY) MON	TUES	WED	THURS	FRI	SAT	SUN	TOTAL QTY THIS PERIOD	UNIT	FOR CONTRACTOR'S USE: UNIT PRICE	COST THIS PERIOD	TOTAL QUANTITIES PREVIOUS	TO DATE	FOR CONTRACTOR'S USE: TOTAL COST PREVIOUS	TO DATE
5.1	Plastic Service by Other than Insertion: Up to 25m Long										ea						
											ea						
											ea						
	Over 25m Long										m						
											m						
											m						
5.2	Plastic Service by Dead Insertion: Up to 25m Long										ea						
											ea						
											ea						
5.3	Plastic Service by Dead Insertion: Over 25m Long										m						
											m						
5.4	Steel Service by Trenching: Up to 25m Long										ea						
											ea						
											ea						
5.5	Steel Service by Trenching: Over 25m Long										m						
											m						
5.7	Above Ground Outside Riser										ea						
											ea						
5.8	Riser Protector										ea						
											ea						
5.9	Excavate, Disconnect, Backfill										ea						
											ea						
6.1	Steel Service: Disconnect, Test, Reconnect										Test						
											Test						
6.2	Plastic Service: Disconnect, Test, Reconnect										Test						
											Test						
	TOTAL FOR THIS INVOICE:																
	HOLD BACK:																
	PAY OUT:																

Figure 11-1. Continued

PROGRESS PAYMENT REPORT

DAILY/WEEKLY

PAGE ___ OF ___
SHEET D: CUSTOMER SERVICE

CONTRACTOR		CONTRACT No.		PLANT No.		W.O. No.		INVOICE No.		PAY PERIOD	
STREET (BLOCK) ADDRESS										DRAWING No.	
PREPARED BY	DATE	ACCEPTED BY	DATE	CHECKED BY Co.	DATE	Co. APPROVAL	DATE	Co. APPROVAL	DATE	Co. APPROVAL	DATE

PAY ITEM No.	DESCRIPTION	SIZE mm	DAILY QUANTITIES (FOR DAILY REPORT ONLY)							TOTAL QTY. THIS PERIOD	UNIT	FOR CONTRACTOR'S USE		TOTAL QUANTITIES		FOR CONTRACTOR'S USE	
			MON	TUES	WED	THURS	FRI	SAT	SUN			UNIT PRICE	COST THIS PERIOD	PREVIOUS	TO DATE	PREVIOUS	TO DATE
																TOTAL COST	
7.1	Remove Meter Set at Building										ea						
7.2	Coring Through Wall, Plug Hole										ea						
7.3	Hang Meter Set										ea						
7.4	Multi-Meter Header Installation										Instl						
7.5	Internal Screwed Piping (Includes Vent Piping)										m						
7.6	Internal Welded Piping (Includes Vent Piping)										m						
7.7	Test and Reconnect Internal Piping										Test						
7.8	Relight Appliances, Do Safety Check										Prem						
7.9	Underground Secondary Line										m						
7.10	Fabricate and Install Riser Return Bend										ea						
	TOTAL FOR THIS INVOICE:																
	HOLD BACK																
	PAY OUT:																

Figure 11-1. Continued

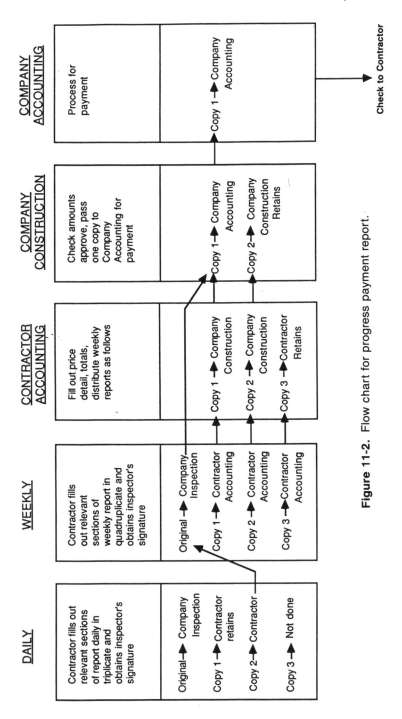

Figure 11-2. Flow chart for progress payment report.

(text continued from page 122)

Notes.

1. The schedule of unit rates and the progress payment reports described in this chapter are suited for direct-burial and dead-insertion work. You will notice that in the case of the schedule of unit rates, item numbers are not consecutive. This has been done to accommodate other pay items such as the ploughing of plastic pipe, etc., which are not applicable to this work. In this way, pay item numbers are consistent throughout the company's various contracts, encompassing its rural as well as urban work. This makes the analysis of costs by category easier, especially if data are inputted directly into a computer.
2. The progress payment report lists only those pay items that are frequently encountered. Provisions have been made by a number of blank rows to list infrequent items, should the need arise.

Benefits

The advantages of using the schedule of unit rates are:

- Competitive bids on a uniform basis can be obtained. Units are explicit.
- Costs for mains, services, and customer-service work can be easily separated.
- Bids can be obtained even before detailed construction plans are ready, i.e., bids can be based on estimated quantities. This method greatly reduces the waiting time (spent in pre-construction engineering) before the actual construction can commence.

Similarly the advantages of the proposed progress payment reporting method are:

- Daily/weekly/monthly records of all construction activities are kept.
- Confirmation of work completed can be done promptly and regularly.
- Confidentiality of contractor's bid prices is not compromised.

12
Criteria for Replacement

INTRODUCTION

These replacement criteria provide guidelines for determining what mains and services should be replaced and which existing pipe should stay and which should be taken out of service. These criteria take the blanket approach of renewing aging distribution piping on a mass scale, to be done on a sequential basis according to urgency. It is assumed that a general evaluation of the distribution system has already been done and that it has been decided that some or most of the system piping needs replacement.

The criteria for replacement are based on the following philosophy:

- Revisits, within 5 years of the completion of the initial replacement program, to make repairs or perform subsequent upgrading will not be in the utility's best interest from a public relations point of view. Thus, all utility appurtenance must be thoroughly reviewed and upgraded the first time.
- Safety is the prime objective. Steel mains or steel services should be grouped together, insulated from customers' piping, and cathodically protected. It will be operationally difficult and uneconomical to monitor isolated single steel services or short pieces of steel mains scattered all over. In such an event, regular cathodic protection monitoring becomes cumbersome and is likely to be missed. Lack of monitoring can cause a safety risk. A steel main should not be upgraded unless a whole block length is involved; similarly, steel services should be at least in one-block groups, or substantially whole blocks.
- Economics dictate that short lengths of steel mains in the run of the PE insertion mains should not be saved.
- Intermixing steel and plastic pipe should be avoided as much as possible. The use of PE material is preferred over steel, as PE neither corrodes nor requires cathodic protection. New steel services off PE mains are disallowed; however, PE services can be laid off the steel mains.

129

CRITERIA FOR REPLACEMENT

Overriding Requirements

1. Steel mains less than one city block long should be replaced by PE mains.
2. Steel services off PE mains, if saved, must be in groups of whole blocks or substantially whole blocks.

Exception: Individual steel services, if connected to steel mains and otherwise qualifying, should be retested to stay in service. For other requirements, see the detailed discussion following.

Detailed Requirements: Mains Replacement

Mains should be replaced with new pipe of suitable size and material according to the criteria given. The method of replacement, i.e., direct burial versus insertion, is also discussed. Pipe should be replaced in the following circumstances:

- If the pipe is bare steel, galvanized, or has screwed joints.
- If it is non-standard size pipe.
- If the pipe is wrapped but it did not have cathodic protection for more than 5 years during its life span. Experience indicates that wrapped steel pipe left in the ground without the cathodic protection for over 5 years develops corrosion pits at the holidays. (Refer to Chapter 3.)
- If the pipe fails the pressure test or it had repair work done to it in the past. A main, if suspected of not passing the pressure test, should be replaced, as it is more costly to pressure test, locate, and fix the leaks than to replace it by insertion. A review of records and previous experiences should be taken into account when considering the main for the upgrading.
- If the pipe is suspected of having been damaged by construction activity or by an unsuitable backfill, such as the roadcrush.
- If it is PVC and PE pipe not conforming to the currently applicable standards.
- If it will be uneconomical or otherwise undesirable to upgrade the existing pipe for the new, higher pressure service. *It is uneconomical to upgrade any main that has more than three unstrapped mechanical couplings per block.* For an analysis of the effect of mechanical couplings on cost, refer to page 137, "Calculations for the Most Economical Length: Upgrading Existing Steel Main."

- If the existing location is unsuitable, i.e., it has shallow cover, a location undesirable for service connections or other future work, or if the present location will require diversion or abandonment in the foreseeable future.
- If the existing pipe fitting ratings are not consistent with new operating pressure.

Maintenance of cathodic protection is important, and it is an added regular cost. The more numerous the CP systems, the more difficult and costly it is to maintain them. Steel pipe requires CP and suitable test facilities for monitoring its effectiveness. In cases where CP monitoring test-post locations cannot be found or steel pipe is in high interference area, e.g., trolly routes or subway routes, or if it is suspected that it may be shorting to other adjacent metallic structures (due to congestion of underground apparatus), steel pipe should be replaced with PE.

Steel pipe, unless it has good coating combined with adequate, continuous cathodic protection, will develop corrosion leaks. Replace steel pipe that has been without cathodic protection for over 5 years. From a maintenance consideration, isolated steel pipe sections should not be any smaller than *one city block length*, as otherwise it will create too many mini-CP systems.

Exceptions: Recently installed PE pipe, if in good condition, should not be replaced. It should be upgraded for higher pressure in accordance with the applicable standards. Recently installed steel pipe should be reviewed for continued service on a case-by-case basis in accordance with the criteria given above.

Detailed Requirements: Services Replacement

All previous criteria for mains apply with the following exceptions:

- Steel services, if connected to steel mains, can be qualified on a singular basis rather than restricted to the one-whole-block requirement for mains, as doing so will not increase the number of CP systems.
- Steel services off PE mains should be grouped in a minimum of one city block, i.e., consider upgrading all the services in a block or replace them all. (CP systems for services will be monitored on a different frequency than for mains.) Exceptions to this rule include when one or more services in a block are less than 5 years old, or when the services are to be connected to steel mains. In this case, less than the whole city block of steel services may be retained. The method of cathodic protection of steel pipe should be specified on the construction drawings.

- PE services may be laid off the steel main. New steel services should not be laid off the PE main.
- PE services may be directly buried, inserted, or partly inserted, part directly buried.
- Steel services 10 m in length or shorter should be replaced unless they are less than 5 years old, in which case they are subject to upgrading. For ease of monitoring, it is preferable to keep the number of separate CP systems to a minimum.
- Replace those service lines requiring a diversion, because it is equally convenient to lay a new service line as it is to do a partial diversion and a partial upgrading. Service lines should not be partially renewed. Replace entire service line.
- Header services should be considered for splitting into individual service lines.
- Service lines not located properly in their easement should be replaced within easement.

HOW TO REPLACE: INSERTION VS. DIRECT BURIAL

Replacement should be by insertion except in the following circumstances:

- If there is route change, i.e., where the existing location is undesirable.
- If there are an excessive number of bell holes rendering the insertion method more expensive. This may arise due to the large number of bends, service connections, and other fittings.
- If the new pipe is steel, direct burial is the only method. Steel pipe should be avoided unless needed for its strength or a code requirement. Insertion of steel pipe in steel pipe is not recommended. However, the use of the steel pipe when required for railway crossings or due to other government regulations should be followed. Wherever the direct burial method is used, the steel pipe should meet the existing standards for such construction, and after completion it should be cathodically protected in accordance with the applicable standard.
- If the existing pipe is too small or restricted for insertion.

Pages 133–136 show the method of calculating the break-even point (distance) below which insertion becomes uneconomical. Since labor and material costs change from time to time, the break-even point will change accordingly. This number should be reviewed once a year after receiving price updates. It should also be noted that the break-even point is calculated based on gas pipe having normal cover and being located in asphalt lanes. Deeper cover will tend to make this distance shorter. Similarly, surfaces such as a gravelled lane where reinstatement costs are low will make this distance longer, whereas a concrete surface for which the excavation and repaving costs are higher will again make the break-even distance shorter.

Other factors that may influence the decision to insert or to directly bury are the number of fittings, terrain, availability of alternate locations for direct burial, and traffic considerations.

The decision whether to insert or to directly bury should initially be made at the planning and design stage and be based on cost information, depth survey, load survey, and other pertinent information. If actual conditions upon excavation are found to be quite different, the company representative can make a decision to suit the circumstances.

CALCULATIONS FOR THE MOST ECONOMICAL LENGTH: OPEN TRENCHING VS. INSERTION

Assumptions

1. Existing steel main is in back lane with lateral service runs on both sides of main. Lane is paved.
2. Normal cover on main.
3. Existing LP main size is 114.3 mm. Replacement PE main size is 60.3 mm.
4. New main, if installed by the direct burial method, can be placed adjacent to the existing main, i.e., within 600 mm of it. New main will cross 50% of the existing services. Some widening of the trench at the service connections will be required.
5. 1987 prices for contract labor, material, and reinstatement are used.
6. All existing services will be reconnected to the new main.
7. Bell hole size for service tee connection shall be no longer than 1 m × 1 m × 1.5 m. This allows sufficient room for cutting away a cylindrical portion of existing main.
8. Cost of bell hole for service by insertion method includes the cost of cutting away cylindrical portion of pipe and subsequent backfilling.
9. Nominal insertion length is 180 m.
10. Job is considered to be of continuous nature, hence only one new bell hole is assumed for each new insertion segment.
11. Overheads are not considered.
12. Cost of live gas operation, i.e., drilling bag holes, bagging, etc., is not included in this analysis, as those costs are the same for both the insertion and the direct burial method.

Data

L = Length of main considered

N = Number of service bell holes

L/N = Distance (m) between service bell holes

B = Cost (per m) of replacement main by direct burial including repaving and material costs
= $99.45/m (contract price)

I = Cost (per m) of insertion. Bell holes for services are not included, but cost of insertion holes, all repaving, and materials are included.
= $35.20/m (contract price)

H = Cost due to additional excavation required for the bell hole at the service tee. This is the cost difference between the service-tee bell hole for insertion and the service-tee bell hole for direct burial. Refer to Figures 12-1 and 12-2.
= $38.10/service (contract price)

Y = Service tee bell hole size for direct burial, 0.6 m × 1 m × 1.5 m per Figure 12-1.

Z = The difference between the bell hole size for service tee in insertion method and the direct burial, i.e., 1 m × 1 m × 1.5 m − 0.6 m × 1 m × 1.5 m = 0.4 m × 1 m × 1.5 m

Therefore the insertion hole size = Y + Z

E = Unit reinstatement cost

Notes:
1. Detailed costs for B, I, and H can be found on pages 139–144.
2. Labor charges for excavating are included in contractor's per-meter pipe-length prices. Fillcreting and asphalting are extra.

Formula

Cost by direct burial = unit direct burial cost × length considered
+ no. of service tee bell holes of size Y × unit reinstatement cost
= B L + N (Y) E

Cost by insertion = no. of bell holes of size (Y + Z) × unit reinstatement cost + unit insertion cost × length considered
= N (Y + Z) E + I L

Equate cost by direct burial to cost by insertion technique

Therefore: $B L + N Y E = N Y E + N Z E + I L$
Therefore: $B L = H N + I L$
where Z E is the additional excavation cost = H
$H N = B L - I L = L (B - I)$
$L/N = H/(B - I)$ Hayat's Equation (2)

i.e., minimum economic distance between service bell holes = $H/(B - I)$

Calculations

Minimum distance between service bell holes:
$= H/(B - I)$
$= 38.10/(76.93 - 17.07) = 38.10/59.86 = 0.59$ m

Conclusions

From an economic point of view, the minimum length needed between service bell holes is 0.59 m. This is the break-even point. It will be economically beneficial to use the insertion method when service tees are more than 0.59 m apart. The reader is advised to take other factors into consideration, and not rely solely on the 0.59 m distance criterion. The following factors should also be considered:

1. Availability of alternate locations
2. Other utility crossings

3. Disruption of traffic
4. Pavement location vs. boulevard location
5. Adequate depth of cover
6. Accessibility for future maintenance
7. Protection given by steel casing
8. Mapping and survey costs associated with new routing

CALCULATIONS FOR THE MOST ECONOMICAL LENGTH: UPGRADING EXISTING STEEL MAIN

Assumptions

1. The method of replacement will be by the dead insertion technique or by direct burial. An economic length is calculated for each alternative.
2. The coating on existing steel pipe is assumed to be good, hence requiring no repair.
3. Cathodic protection will be provided by anodes. Based on 114.3-mm steel pipe, an anode every 500 m is assumed. The cost of anode plus test post is assumed to be $200 per installation. This cost does not include excavation, as it is assumed that bell holes will be available due to other activities, such as for the pressure test.
4. Main is assumed to be in asphalted back lane.
5. New pipe is 60.3-mm PE, replacing 114.3-mm steel pipe.
6. Mechanical couplings need strapping.
7. Maintenance costs for CP system are negligible. Costs of isolating services at buildings are ignored.
8. Retesting of main and associated services can be done in one test. No separate bell holes are needed for service testing, and 100% of tests will pass in the first attempt. Cost of insertion, I, does not include any amount for the service bell holes as they are not considered in this comparison.
9. Overheads are not considered.
10. Cost of live gas operation, i.e., drilling bag holes, bagging, etc., is not included in this analysis.

Data

Cost of insertion	I = \$35.20/m	Contract price includes materials and repaving.
Cost of retesting including bell holes	R = \$ 7.65/m	Contract price.
Cost of strapping mechanical coupling	D = \$1500.00	Price includes material and repaving.
Cost of anode and test post installation	A = \$200.00 ea	Price does not include excavation and repaving.
Cost of replacement by direct burial	B = \$99.45/m	Contract price includes material and repaving.

$$N' = \text{no. of mechanical couplings}$$

$$L = \text{length in meters}$$

Assume an anode and test post installation for 114.3-mm main every 500 m.

Note: Calculation of R is detailed on page 142. For details on I and B, refer to pages 139–140.

Formula

Cost of replacement = Cost of retesting existing main and cost of installing CP system.

For replacement by insertion:

$$I\,L = R\,L + D\,N' + \frac{L}{500} \times A \qquad \text{Hayat's Equation} \quad (3)$$

For replacement by direct burial:

$$B\,L = R\,L + D\,N' + \frac{L}{500} \times A \qquad \text{Hayat's Equation} \quad (4)$$

If no mechanical coupling, then:

$$B\,L = R\,L + \frac{LA}{500} \text{ or } B = R + \frac{A}{500}$$

Since B is 99.45, R is 7.65, and A is 200, the cost of direct burial will always be greater than retesting.

Similarly, I = R + A/500. Since I = 35.20, R = 7.65, and A = 200, the cost of insertion when no unstrapped mechanical coupling is involved will always be greater than the retesting cost.

Calculations

For replacement by insertion:

$$I\,L = R\,L + D\,N' + \frac{AL}{500}$$

$$35.20 \times L = 7.65\,L + 1{,}500\,N' + \frac{200}{500}\,L$$

$$1{,}500\,N' = 35.20\,L - 7.65\,L - 0.4\,L = 27.31\,L$$

$$N' = \frac{27.3\,L}{1{,}500} = 1.8 \times 10^{-2}\,L$$

Per block (180 m) $N' = 1.8 \times 10^{-2} \times 180 = 3.3$

For replacement by direct burial:

$$B\,L = R\,L + D\,N' + \frac{AL}{500}$$

$$99.45 \times L = 7.65\,L + 1{,}500\,N' + \frac{200\,L}{500}$$

$$1{,}500\,N' = 99.45\,L - 7.65\,L - 0.4\,L = 91.4\,L$$

$$N' = \frac{91.4 \text{ L}}{1,500} = 6.09 \times 10^{-2} \text{ L}$$

Per block (180 m) $N' = 6.09 \times 10^{-2} \times 180 = 10.96$

Conclusions

In a typical city block (180 m), it will be more economical to replace the main than to upgrade it by retesting and providing CP when the number of mechanical couplings is *4 or more for dead insertion* and *11 or more for direct burial*.

In other words, insertion will be cheaper than upgrading a steel main as long as 4 or more couplings per block require strapping. Similarly, direct burial will be cheaper than upgrading a steel main as long as 11 or more Dressers per block require strapping. Before deciding in favor of retesting existing pipe whose integrity may be in question, also consider the maintenance cost of a CP system and the interference problems from stray currents generated by the electric subway and bus systems. From a maintenance point of view, CP systems should not be smaller than one-whole-block length.

Also consider the condition of main. If it fails the pressure test, it may become quite expensive to locate and repair the leak and to retest it. The cost of retesting (R = 7.65/m), shown in the above calculations, is based on main passing the pressure test at the first attempt.

Calculation of B

B = Cost of replacement main per meter by direct burial

Consider a 60.3-mm PE main installed in an open trench adjacent to the existing main, or within 1 m of the old main. With 1.2 m average cover, the trench dimensions will be 0.6 m wide × 1.5 m deep × 180 m long (one whole block). Assume the location is an asphalted back lane. B has three components: labor, reinstatement, and material.

Labor (L)

Laying 60.3-mm PE	= $23.75/m
Soil removal and fillcrete placement (cost included in labor)	= 0

Asphalt removal: 0.6 m × 1.5 m × 1 m
$$= 0.9 \text{ m}^3/\text{m at } \$23.73/\text{m}^2$$
$$= 0.9 \times 21.73 \qquad = 19.60/\text{m}$$

Removal of old 114.3-mm steel pipe	= 8.00/m
Total	= $51.35/m

Reinstatement (R′)

Backfilling with fillcrete and asphalting. (The trench will be backfilled using fillcrete up to the last 0.15 m, which is filled with asphalt.)

Fillcrete	$30/m^3 × 0.6 × 1.35 × 1 m^3/m	= $24.30/m
Asphalt placement	$33/m^2 × 0.6 × 1 m^2/m	= 19.80/m
Total		= $44.10/m

Material (M)

60.3-mm pipe and fittings = $4.00/m

Total B

B = L + R′ + M = $51.35 + 44.10 + 4.00

Total	= $99.45/m

Calculation of I

I = Cost of insertion per meter

Assuming a 60.3-mm PE main to be installed inside a 114.3-mm steel casing. The PE sleeve size shall be 88.9 mm. Assume continuous work, thus only one insertion hole attributed to each block. There are three cost components: labor, reinstatement, and material.

Labor (L)

Inserting 60.3-mm PE carrier and 88.3-mm sleeve

60.3-mm PE @ $22.75/m
88.3-mm sleeve @ 4.50/m

Total = $27.25/m

Reinstatement (R′)

One insertion hole per block. Hole dimensions are 2.4 m long × 1 m wide × 1.5 m deep.

Asphalt removal	$= \dfrac{2.4 \times 1}{180}$	$= 0.013 \text{ m}^2/\text{m}$
	$= 0.013\text{m}^2/\text{m} \times \$21.73/\text{m}^2$	$= \$0.28/\text{m}$
Asphalt replacement	$= 0.13 \text{ m}^2/\text{m} \times \$30/\text{m}^2$	$= 0.43/\text{m}$
Fillcrete	$= \dfrac{2.4 \times 1 \times 1.35 \text{ m}^3}{180 \text{ m}} \times \$30/\text{m}^3$	$= 0.54/\text{m}$

Total   = $1.25/m

Material (M)

Pipe and fittings = $4.00/m

Sleeve   = 2.70/m
————————————————
Total   = $6.70/m

Total I

I = L + R′ + M = $27.25
                            1.25
                            6.70
————————————————
Total   = $35.20

Calculation of R

R = Cost of retesting mains and services per meter

It is assumed that one whole block length (180 m) of 114.3-mm steel main contains 30 existing services. Test bell holes for the main will be approximately half the size of the hole required for the insertion bell hole. A test hole at each end of the block will be required.

Therefore, excavation required = $2 \times 1.2 \times 1 \times 1.5$ m^3

There are two cost components: labor and reinstatement. The labor costs include charges for excavation, disconnection, testing, purging, reconnecting, and backfilling. In this case, the labor costs have been set by contract to be $1,200.

The reinstatement costs are calculated as follows:

Asphalt = $2 \times 1.2 \times 1$ m^2 $\times$ $33/m^2 = $79.20

Fillcrete = $2 \times 1.2 \times 1 \times 1.35$ m^3 $\times$ $30/m^3 = $97.20

Therefore, the total cost per 180-m block   = $1,200.00 + 79.20 + 97.20
$$\text{Total} = \$1,376.40$$

$$R = \frac{\$1,376.40}{180} = \$7.65/m$$

Calculation of H

H = Cost for additional excavation for the service bell hole, i.e., the difference between bell hole cost for insertion and direct burial.

For insertion, service tee bell hole has to be slightly larger. The service tee bell hole size for the direct burial is 0.6 m $\times$ 1 m $\times$ 1.5 m (Figure 12-1), and

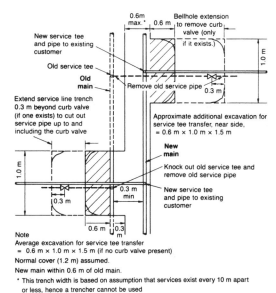

Figure 12-1. Typical bell hole size for service transfer, main paralleled.

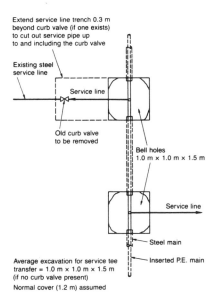

Figure 12-2. Typical bell hole size for service transfer, main inserted.

for insertion it is 1 m × 1 m × 1.5 m (Figure 12-2). Therefore, the additional excavation per service connection is 0.4 m × 1 m × 1.5 m. The following cost components apply:

Asphalt removal	$21.73/m^2 × 0.4 m^2 =	$ 8.70
Spoil removal	included =	0.00
Placing fillcrete	$30/m^3 × (0.4 × 1 × 1.35) m^3 =	16.20
Placing asphalt	$33/m^2 × 0.4 m^2 =	13.20
Total H		= $38.10

13
Dead Insertion

INTRODUCTION

Planning an insertion operation should start well in advance of the actual installation. The procedure given in this chapter should be taken as a recommendation only, and must be modified in the light of prevailing conditions, regulations, and local preferences.

DESIGN CONSIDERATIONS

Mains to be Replaced

Deciding the method of replacement and which pipe to replace may be determined on the basis of information given in Chapter 12. The insertion technique should not be considered for any existing main with less than 600-mm (24-in.) cover or service with less than 450-mm (18-in.) cover, i.e., if the cover is less than the minimum allowed by code. Similarly, insertion in abnormally deep mains (over 1.5-m cover) must be reviewed by the company engineer.

Gas Flow

A main or service inserted with PE will have a lower flow capacity than the original main or service, unless the operating pressure is elevated adequately to compensate for the reduced flow area. Sizing done by network analysis will take all pertinent factors into consideration. If hand calculations are done, similar considerations must apply.

Field and Record Survey

A records investigation and field survey should be conducted to accurately determine the sizes, location of the mains, offsets, tees, valves, drips, all services, and service entrances. All of this information should be marked on construction drawings for use while locating and fabricating the system. This information should be confirmed by line locations before any actual excavation commences.

Carrier Pipe Size for Insertion

PE carrier pipe sizes for low pressure mains to be converted to MP (105 kPa MOP) are given in Table 13-1, which may be used as a rough guide for sizing. For the replacement of IP and MP mains, size determination must be done by network analysis. Similarly, replacement sizes for low-pressure mains should be found using a network analysis program. Results of network analysis may give different sizes than those shown in Table 13-1. If sizes arrived at by network analysis are smaller than those recommended in Table 13-1, then use the recommended sizes for insertion shown in the table. Compliance with this requirement is necessary in order to avoid a possible freeze-off and squeezing of PE pipe in the steel casing. An alternative to this would be to use the calculated smaller sizes, but use the direct burial method of construction, or, if the insertion method is used, to fill the casing with a suitable filler, such as a sleeve pipe, foam, or foam chips.

An economic analysis should be done to evaluate each alternative. A useful consideration is that, for the same size pipe, the replacement cost by insertion is 50 to 70% of the direct burial cost. Where the calculated size is larger than the allowable carrier size given in Table 13-1, the insertion method is not applicable, hence the direct burial method must be used.

The sizes given in Table 13-1 have been determined after due consideration to the flow capacities and the effect of water freezing within the steel casing. The sizes given will provide equal or greater flow than that of the original casing pipe operating at 1.72 kPa pressure.

Sizing the PE carrier pipe is based on an operating pressure of 105 kPa, using SDR 11 PE pipe (except for the 42.2-mm (1¼-in.) size, which is SDR 10). For a detailed discussion of how carrier pipe sizes were determined, refer to Chapter 4. The calculations on the effects of freezing water in the annulus have been confirmed by a series of tests described in Chapter 6.

Table 13-1
Pipe Replacement Sizes for Insertion

LP Steel Casing Size	Allowable PE Carrier Size for Insertion
26.7 mm (¾ in.)	15.9 mm (½ in. CTS)
33.4 mm (1 in.)	26.7 mm (¾ in.)
42.2 mm (1¼ in.)	26.7 mm (¾ in.)
48.3 mm (1½ in.)	26.7 mm (¾ in.)
60.3 mm (2 in.)	42.2 mm (1¼ in.)
88.9 mm (3 in.)	* 42.2 mm (1¼ in.)
	60.3 mm (2 in.)
114.3 mm (4 in.)	* 60.3 mm (2 in.)
	# 88.9 mm (3 in.)
168.3 mm (6 in.)	* 88.9 mm (3 in.)
	114.3 mm (4 in.)
219.1 mm (8 in.)	*114.3 mm (4 in.)
	168.3 mm (6 in.)
273.1 mm (10 in.)	168.3 mm (6 in.)
	219.1 mm (8 in.)
323.9 mm (12 in.)	*168.3 mm (6 in.)
	219.1 mm (8 in.)

* Sleeve pipe is mandatory for these combinations due to freezing considerations. As sleeving adds to the costs, these combinations should be avoided. However, when necessary, a sleeve or some other method of protection against external scratching of PE carrier pipe may be adopted for all inserted pipe as per the company's specifications and plans. If a sleeve is specified, no other alternative method of scratch protection is required.

This combination is not recommended due to insufficient clearance. Neither a sleeve nor a thinsulator can be accommodated in this case. Steel pipe must be cleaned and deburred properly before insertion.

Meter Sets

Upgrading of LP (1.72 kPa) Mains and Services

1. Residential Meter Sets
   Loads less than 28 m³/hr (1,000 scfh)
   - Alter interior piping, complete with coring.
   - Relocate meter from inside to outside, complete with meter change card.
   - Install a suitable regulator upstream of the meter. If necessary, install the riser protector.
   - Complete the safety survey.
   - Complete the dead check.
   - Complete the appliance relight.

2. Commercial/Industrial Meter Sets
   Loads up to 144 m³/hr (5,000 scfh)
   - Meter may generally be located inside while the regulator is installed outside on the aboveground riser. A riser protector should be installed as per the company's specifications.
   Loads of 144 m³/hr to 314 m³/hr (5,000 to 11,000 scfh)
   - Both meter and regulator should generally be located inside the premises. Relief valve and regulators must vent to the outside. For an IP set, a suitable meter room should be provided.
   Loads above 314 m³/hr (above 11,000 scfh)
   - Same instructions as for loads of 144 m³/hr to 314 m³/hr. Aboveground meter or riser subject to accidental damage should be protected with a riser protector.

Upgrading of MP (55 kPa or 105 kPa) Mains and Services

Follow the same instructions as for LP upgrade.

Upgrading of IP (550 kPa) Mains and Services

Follow the same instructions as for LP, noting that urban residential customers should not be supplied off the IP mains. Consideration should be given to transferring existing commercial and industrial customers off the IP system onto the MP system. Only customers with high gas-consumption needs (> 3,000 m³/hr) should be considered for servicing from the IP system.

Note that all services must have an outside aboveground riser (Figures 13-1 and 13-2). A below-grade service entry without an outside shut-off valve at the building may, in some jurisdictions, require government approval.

CONSTRUCTION CONSIDERATIONS

Insertion

Before the actual construction begins:

- Obtain an excavation permit where required by regulation.
- Arrange road barriers and other necessary traffic control equipment.
- Ensure adequate manpower availability.
- Arrange hot tapping and stopping equipment and necessary pipe, fittings, meters regulators, etc.

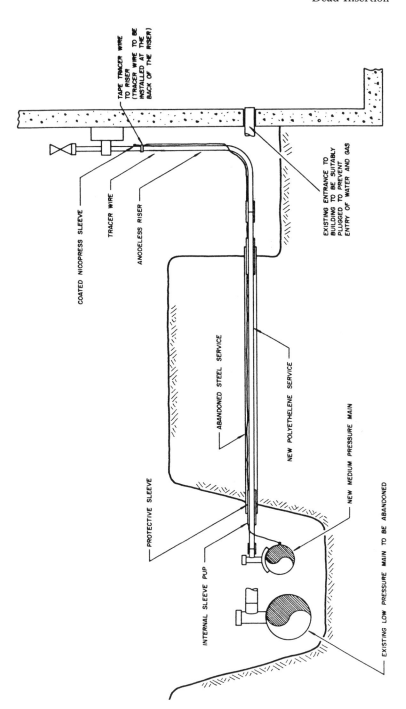

Figure 13-1. Service insertion—PE service off new PE main, aboveground entry.

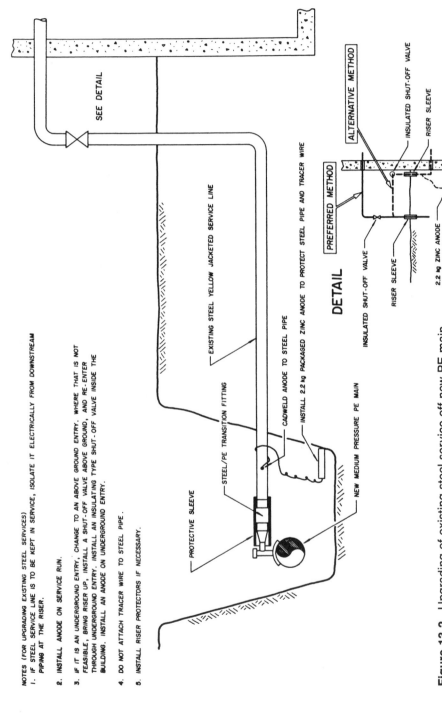

NOTES (FOR UPGRADING EXISTING STEEL SERVICES)

1. IF STEEL SERVICE LINE IS TO BE KEPT IN SERVICE, ISOLATE IT ELECTRICALLY FROM DOWNSTREAM PIPING AT THE RISER.

2. INSTALL ANODE ON SERVICE RUN.

3. IF IT IS AN UNDERGROUND ENTRY, CHANGE TO AN ABOVE GROUND ENTRY, WHERE THAT IS NOT FEASIBLE, BRING RISER UP, INSTALL A SHUT-OFF VALVE ABOVE GROUND, AND RE-ENTER THROUGH UNDERGROUND ENTRY. INSTALL AN INSULATING TYPE SHUT-OFF VALVE INSIDE THE BUILDING. INSTALL AN ANODE ON UNDERGROUND ENTRY.

4. DO NOT ATTACH TRACER WIRE TO STEEL PIPE.

5. INSTALL RISER PROTECTORS IF NECESSARY.

SEE DETAIL

EXISTING STEEL YELLOW JACKETED SERVICE LINE

CADWELD ANODE TO STEEL PIPE

INSTALL 2.2 kg PACKAGED ZINC ANODE TO PROTECT STEEL PIPE AND TRACER WIRE

NEW MEDIUM PRESSURE PE MAIN

STEEL/PE TRANSITION FITTING

PROTECTIVE SLEEVE

DETAIL

PREFERRED METHOD

ALTERNATIVE METHOD

INSULATED SHUT-OFF VALVE

RISER SLEEVE

INSULATED SHUT-OFF VALVE

RISER SLEEVE

2.2 kg ZINC ANODE

Figure 13-2. Upgrading of existing steel service off new PE main.

- Review plans and procedures for shutting off, testing, tying in, purging, and relighting. Ensure that pressure testing and purging equipment is available. Pre-plan interior piping alterations, coring of new entries and vents, and prefabricating and testing return bend piping as required.
- Notify customers of road/alley closures and of the gas shutdown and make arrangements for access to customers' premises. Ensure temporary gas supply where required.
- Obtain foreign facility locations and stake out gas mains, services, any insertion obstructions such as tees, valves, elbows, or reducers and all other utilities in the immediate work area.

Once the number and locations of services are known, the length of section of main which can be inserted in one day can be determined. This is usually limited by the number of services in each section. Typically, this section should not be longer than one city block, or 30 services. The time requirement for testing the main must also be taken into account. Adequate customer service crews must be available to reset the meter/regulator, complete the dead checks, relight customers, and complete safety surveys the same day as the new mains and services are energized and the old carrier pipe is abandoned.

The meter and regulator set should be prefabricated. If it is being relocated from inside to outside, it may be installed prior to the day of the job, ready for connection on the TO day (turn-off day).

A typical weekly construction work schedule is as follows:

One to two weeks prior to the actual construction, deliver advance notices of interior work and street closures to the customers.

Day 1—Move to job site and excavate bell holes on mains and services for insertion, retest, or direct burial.

Day 2—Weld on appropriate fittings for bagging off, purging, and tying in. Make connections for temporary gas supply where required.

Day 3—Disconnect customers, shut off gas sources, and pig the main, ensuring that no damage occurs to existing meters, appliance regulators, etc., due to overpressuring. Purge the gas, and prepare the main and services for upgrading (i.e., direct burial, insertion, retesting, or retirement, whichever is applicable). Carry out upgrading. Install regulators and meters. Dead-test meter and interior piping. Inaccessible premises must be cut off and tagged. Purge air, and relight customers following the safety survey. Make arrangements to visit customers who could not be upgraded or relit this day.

Day 4—Revisit remaining premises and complete the upgrading. Some premises may have to be left cut off from the gas supply until the customers provide access. Backfill all holes and restore the area to its original or better condition.

Day 5—Clean up, move to the next site.

Detailed Steps

1. The local municipal authority regulates traffic flow and restricts construction to certain times. These instructions are issued with the excavation permit. Follow these instructions and all other applicable occupational safety rules.

2. Expose both ends of the main to be inserted (see Figures 13-3 through 13-8). An obstruction such as a valve, bend, or a reducer will require excavation and cutting out. At the ends of the section to be inserted, make the excavation large enough to allow easy entry of the plastic pipe without exerting excessive stress on it. Recommended bell hole sizes for the main insertion hole and the main receiving hole are given in Table 13-2.

Figure 13-3. Asphalt marked for service riser bell hole.

Figure 13-4. Asphalt cutting using a diamond saw.

Figure 13-5. Main and service trenches ready for direct burial.

Figure 13-6. Insertion method—service line bell holes at main.

Table 13-2
Recommended Bell Hole Sizes for Insertion

PE Pipe Size	Insertion Hole L × W (m)	Receiving Hole L × W (m)
15.9 mm (½ in. CTS)	1.5 × 1.0	1.5 × 1.0
26.7 mm (¾ in.)	1.5 × 1.0	1.5 × 1.0
42.2 mm (1¼ in.)	1.5 × 1.0	1.5 × 1.0
60.3 mm (2 in.)	2.4 × 1.0	1.5 × 1.0
88.9 mm (3 in.)	3.5 × 1.0	2.0 × 1.0
114.3 mm (4 in.)	4.8 × 1.5	2.0 × 1.0
168.3 mm (6 in.)	7.2 × 2.0	3.0 × 1.2
218.1 mm (8 in.)	9.6 × 2.0	3.0 × 1.2

Note: The bell hole sizes given are suitable for a depth cover of 1.2 m. Where a main is buried deeper, longer insertion holes are required to avoid bending the pipe beyond its recommended bending radius. Sufficient straight length of PE pipe is required at the entrance for the PE pipe to slide easily into the steel casing. For depths of cover other than 1.5 m, the following formula can be used to calculate the insertion hole length. The width of the insertion hole and both length and width of the receiving hole remain the same as given in the above table.

The length of insertion hole (m) = depth of cover (m) × pipe size (NPS).

Example: NPS 2 (60.3-mm) pipe, 2-m cover

Length of insertion hole = 2 × 2 = 4 m

For bell holes deeper than 1.5 m, provide back slope and shoring.

3. Expose all service tie-ins, service entrances, valves, elbows, drips, and other possible obstructions. Expose all locations of laterals. Protect the excavations with barricades.

4. Make service tie-in excavations the standard size used for repairs and other work, providing enough working room for the service insertions and the tie-ins. Sizes given for the receiving holes in Table 13-2 should be used as a guide.

5. Pre-install any new services.

6. Where full traffic is to be maintained, place steel plates capable of supporting the traffic over the excavations. Provide cold-mix ramping around steel plates. Provide and maintain flashers where necessary. When working in alleys, provide fencing on both sides of the trench while trenches and holes are open.

7. Weld and drill the required fittings on the main in accordance with the company's tie-in and purging procedures (Figures 13-7 and 13-8). Mains and services should be readied for shutoff at this point. Tapping of lines should be arranged.

8. Verify that customers' premises affected by the interruption of gas services will be accessible all day.

Figure 13-7. PE main fused above ground and ready for insertion.

Figure 13-8. PE carrier pipe being moved into position for insertion.

9. Place barricades, detour signs, parking, and no-parking signs as necessary.
10. Organize all tools and equipment, i.e., generators, drills, tools, compressors, flow-stopping equipment, purging equipment, and nitrogen. (See Figure 13-9.) Temporary gas supplies, fire extinguishers, etc., should be checked and readied for the job.
11. At about 8:00 a.m. (after most customers will be finished with breakfast), shut off gas to the segment of main that is to be upgraded. Purge gas from main and each service.
12. Cut the services back a minimum of 1.5 m (5 ft) from the main. Cut out any shutoff valve (curb stop) through which the PE pipe service will not be able to pass. Wherever steel service pipe is cut, its ends must be reamed smooth before inserting the carrier pipe. An alternative to reaming is to use a suitable size sleeve pup 600-mm long, fitted at the ends. A protective sleeve shall be fitted over the steel pipe ends to provide shear protection to the PE carrier pipe.

    Note: Install a jumper cable before cutting out any portion of steel main or service line, or before dismantling the old meter.
13. Curb valves on residential services should be removed since they will be replaced by the stopcocks on the outside risers. Follow construction drawings for location, type, and size of new valves on mains and services. In those circumstances where an outside riser cannot be provided, a curb valve should be installed on the service line.

A PE service line must have an above-grade, outside steel riser and a steel-pipe entry, as an entry to a building with PE pipe is not permitted.

14. Cut back services about 0.6 m (2 ft) from the premises to insert the PE pipe and connect the riser and the meter set. Connect the inside piping to the new meter/regulator set, which will be located outside.

    Note: Install a jumper cable before cutting out any portion of the steel main or the service line. Only cold cutting is allowed unless the main has been pigged, purged, and tested with a combustible gas indicator prior to using a torch.

15. Cut out a cylindrical section of steel main at a point where the connection to the PE main will be made. Cutout lengths are given in Tables 13-3 and 13-4. (Also see Figures 13-10 and 13-11.)

16. Inspect the old main to determine the degree of cleaning required. All mains 60.3 mm (2 in.) and larger should be considered for pigging by a poly pig or by a circular brush or a similar device. To avoid damage to the connected equipment, i.e., the regulators, meters, etc., they should be isolated before pigging. When a main is cut up into small

Figure 13-9. Some commonly needed small equipment. Note the compressed natural gas trailer in the background. It is used to keep supplying non-interruptable gas customers while dead insertion work is in progress.

Table 13-3
Cutout Length of Main for Service Tees, Laterals at Receiving Hole

PE Main Size	Steel Pipe Cutout Length (m)		
	Service Tee	Laterals	Receiving Hole
42.2 mm (1¼ in.)	0.9	1.4	1.4
60.3 mm (2 in.)	0.9	1.4	1.4
88.9 mm (3 in.)	0.9	1.9	1.9
114.3 mm (4 in.)	0.9	1.9	1.9
168.3 mm (6 in.)	0.9	2.9	2.9
219.1 mm (8 in.)	0.9	2.9	2.9

Table 13-4
Cutout Lengths of Main for Insertion at Insertion Hole

PE Main Size	Steel Pipe Cutout Length
42.2 mm (1¼ in.)	1.4 m
60.3 mm (2 in.)	2.3 m
88.9 mm (3 in.)	3.5 m
114.3 mm (4 in.)	4.7 m
168.3 mm (6 in.)	7.1 m
219.1 mm (8 in.)	9.5 m

segments because of the need to install frequent service connections, etc. (as is the case in a residential area), it is not necessary to run the pig. However, each segment should be visually inspected to determine the need and method of cleaning. Obstructions must be removed.

In cases where a pig is run, a graduated nylon rope should be attached to the poly pig before the pig is sent into the steel casing. The graduations on the rope should be read to evaluate the distance of any obstruction encountered from the insertion point. Mains and services smaller than 60.3 mm (2 in.) should be blown clean with air or nitrogen to remove the internal loose material. Sizes 60.3 mm (2 in.) and smaller need not be pigged.

17. Obstructions should be located and cut out of the main. Some mains may require a re-run of the pig to remove all condensate and other extraneous material from the main. The company representative should determine the need to re-run the pig.

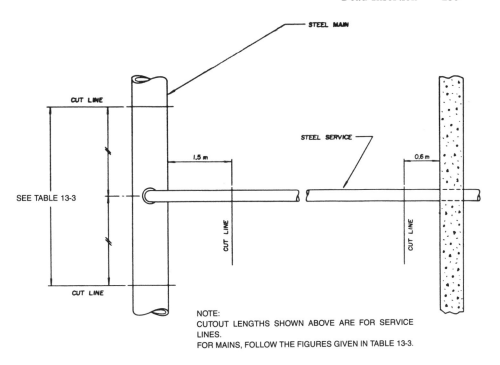

Figure 13-10. Service cutout lengths for insertion.

18. Injuries to the PE pipe during insertion should be kept to a minimum. Depth of scratches must not exceed 10% of the wall thickness. Before inserting any PE pipe 42.2 mm or larger, pull a 4-m-long PE test piece through the steel main using a rope to pull the test piece. Refer to Figure 13-12 for instructions on pulling the test piece.

19. An unacceptable level of damage to the test piece will warrant one of the following actions to be taken prior to the insertion of the carrier pipe. The company representative on site should determine which alternative must be undertaken. The first alternative is the most preferred.

   (a) Lining the steel main with a plastic sleeve. See Table 13-5 for sleeve-dimension information. The sleeve may be inserted by pulling, pushing, or both.

(Text continued on page 163)

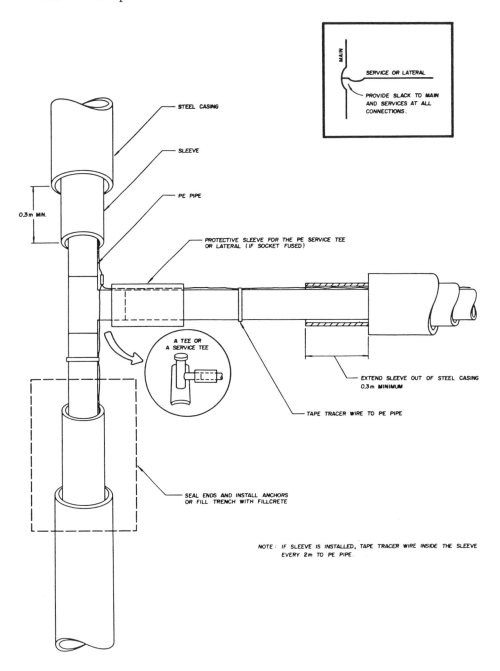

Figure 13-11. Steel main cutout details for lateral or service line.

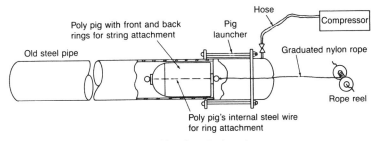

Pig launching with string attachment

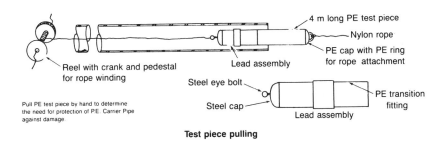

Test piece pulling

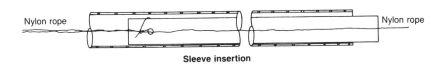

Sleeve insertion

Attach rope to thin sleeve by drilling 2–15 mm holes diametrically opposite to each other. Thin sleeve may be pushed or both pushed and pulled. If thin sleeve is not used, attach rope directly to lead assembly which is to be fused to the PE carrier pipe. PE carrier pipe must be pushed by hand and may be assisted by pulling the rope by hand. The use of the rope pull is optional. Sizes NPS3 and smaller do not need pulling, but it is preferable to fuse the lead assembly or a straight 1.5 m long pipe at the leading end to avoid this end curling up during insertion.

Figure 13-12. Steps of pig launching to carrier pipe insertion.

(Figure 13-12 continued on next page)

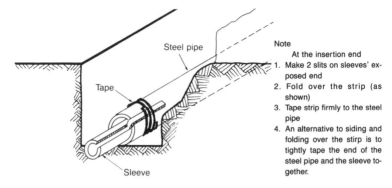

Note

At the insertion end

1. Make 2 slits on sleeves' exposed end
2. Fold over the strip (as shown)
3. Tape strip firmly to the steel pipe
4. An alternative to siding and folding over the stirp is to tightly tape the end of the steel pipe and the sleeve together.

Anchoring sleeve prior to carrier pipe insertion

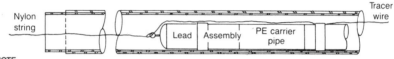

NOTE

Mark PE pipe every 5 m to aid in the location of any obstruction.
If a sleeve is used, it should be marked rather than the PE carrier pipe.

PE carrier pipe insertion

Figure 13-12. Continued.

Table 13-5
PE Sleeve Sizes

Steel Pipe Sizes			PE Carrier Pipe Size		Sleeve	
OD (mm)	NPS (in.)	ID (mm)	OD (mm)	NPS (in.)	OD (mm)	WT (mm)
33.4	1	26.6	15.9	1/2 CTS	21	1.5
42.2	1 1/4	35.1	26.7	3/4	31	1.5
60.3	2	52.5	42.2	1 1/4	47	1.5
88.9	3	82.55	42.2	1 1/4	60.3	3.8
114.3	4	107.95	60.3	2	88.9	3.18
168.3	6	160.3	88.9	3	136.65	4.5
219.1	8	209.5	114.3	4	190.0	6.0
			168.3	6	190.0	6.0
273.1	10	263.4	168.3	6	219.1	3.18
			219.1	8	250.0	5.7
323.9	12	312.67	168.3	6	250.0	10.7
			219.1	8	250.0	5.7
304.8	12 in. OD*	292.1	168.3	6	250.0	5.7

Note: For steel pipe sizes 42.2 mm (1 1/4 in.) and 60.3 mm (2 in.), sleeving for the entire length of steel pipe is not required. For these two sizes, use sleeve pups at ends only.
* Non-standard pipe, used by some companies in the past.

(Text continued from page 159)

(b) Installation of Thinsulators or PE carrier pipe. See Figure 13-13 for Thinsulator installation.

(c) Use of thicker-wall carrier pipe, i.e., SDR 8.8 PE instead of SDR 11.

Note: This alternative cannot be decided on the spot. It has to be evaluated, and accordingly, pipe delivery must be arranged well in advance of the job commencement.

(d) Run a steel pig through the line to blunt and remove sharp corners. The pig may have to be run more than once, preferably back and forth, until the damage to the test piece is observed to be less than 10% of the PE pipe thickness.

20. If a sleeve is not used, the ends of old steel pipe must be fitted with the sleeve pups to protect the carrier pipe against damage. See Figure 13-14 for sleeve-pup installation.

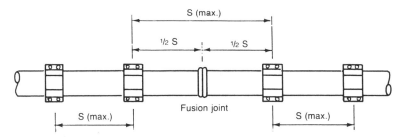

SPACING OF THINSULATORS—S (max.) - meter

Type of Pipe	Size of PE pipe					
	60.3	88.9	114.3	168.3	219.1	273.1
PE 3406-SDR 11	2.0	2.5	2.5	3	3	3.5
PE 2306-SDR 11	2.0	2.0	2.5	2.5	2.5	3.0

NOTES: 1. Thinsulator size will be specified for each job. It will depend on the size of casing pipe and the size of PE pipe and on what obstructions are suspected in the casing. The decision to use Thinsulators or not will be made by the engineer on an individual job basis. For socket-fused pipe, no Thinsulator is needed.
2. Spacing chart given above is for PE 3406/2306 SDR 11 pipe. For other pipe, consult the engineering department.
3. Spacing S may be reduced to maintain approximately equal spacing.
4. Thinsulator must be tightened properly so that it does not move during pushing operation.

Figure 13-13. Thinsulator installation on PE pipe for insertion in steel casing.

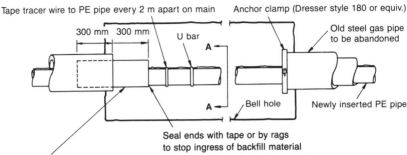

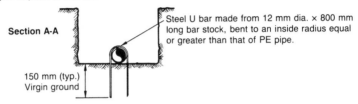

Note:
1. For Dresser anchor clamp installation, cut sleeve to within 10 mm of casing end and anchor PE pipe near end of casing.
2. If gas line is in travelled section of street, trench may be backfilled using fillcrete, in which case anchors are not necessary (as fillcrete provides anchoring against axial movement).
3. Anchor clamps are required if bell holes are more than 15 m apart and backfill is soft and sandy soil, not providing natural ground grip around PE pipe.

Figure 13-14. Anchor and sleeve pup installation for plastic main insertion.

21. Insert PE carrier pipe with the tracer wire attached to it. Tape tracer wire (AWG 14 TWU copper) to pipe at 2-m intervals and provide slack between each taped length. The leading end of the carrier pipe should be fitted with a lead-assembly. For details of the lead-assembly, see Figure 13-12.

    The PE carrier pipe may be inserted by pushing (Figure 13-15). Hand pushing assisted by mechanical equipment while holding the pipe on fabric slings is permitted. The leading end of the carrier pipe may be assisted by an operator maintaining a continuous tension on a rope attached to the leading end of the carrier pipe. Where the steel main is on an incline, use the high point for the insertion hole.

22. If the PE pipe hangs up during insertion, withdraw pipe a meter or so, rotate pipe 90° and re-push. If the obstruction appears to be solid, it may be an elbow, a mechanical coupling, weld icicles, a valve that was not recorded, a dented pipe, or a similar obstruction. Locate, excavate, and cut out the obstruction and resume insertion.

23. After the insertion is complete, run a poly pig through the new main, then install end caps, and test the main according to the test schedule given in Table 13-6. (See Figures 13-16 through 13-18.)

24. While the test on the main is in progress, insert service lines. Inspect the inside of the old service pipe. Remove burr by hand-reaming the steel service pipe ends with a tapered reamer. If necessary, blow internal rust out using compressed air or nitrogen. Fit a nose cone (Figure 13-16) or a cap on the leading end of the PE pipe to stop extraneous material entering the pipe. Copper tracer wire AWG No. 16 TWU is to be taped to the PE pipe at 2-m intervals before it is inserted. Ensure that there is slack in the wire between each taped length.

Table 13-6
Pressure Test Requirements

Pipe Length Mains	Test Pressure	Duration of Test	Medium
Less than 100 m	770–1,100 kPag	15 minutes	Air/nitrogen
100 m–300 m	770–1,100 kPag	4 hours	Air/nitrogen
300 m–5,000 m	770–1,100 kPag	12 hours	Air/nitrogen
Over 5,000 m	770–1,100 kPag	24 hours	Air/nitrogen
Services			
Less than 100 m	770 kPag	10 minutes*	Air/nitrogen
For services 100 m and longer, follow instructions for mains.			

* Ten minutes if a bubble-type testing instrument is used; otherwise test for fifteen minutes using a pressure gauge.

Figure 13-15. PE carrier pipe being pushed into the steel main.

25. Push through enough PE service pipe to tie-in with the main. Cut off enough length to make connection to the service tee and to the riser. Provide enough slack to the service line at the service connection to the main, and ensure that the new PE pipe rests on compacted soil. Assemble the riser at an offset and elevation as required to properly set the new meter and regulator. (See Figures 13-19 through 13-23.)

26. If the pressure test shows no leaks, depressurize the main, make the connection to the live gas source and purge the air. Follow company's tie-in and purging procedures.

27. Prior to or after the main is gassed up, fuse each service tee on the main. A protective sleeve should be installed over the outlets of the service tees before each service is fused to the tee. Similarly, a protective sleeve pup is required at each entry of the cut-and-reamed steel service pipe. Cut the services to the right length and fuse into the tapping tees.

NOTE :

1. FOR INSTALLING NOSE CONE, SLIP RUBBER STUB IN THE PE PIPE END AND HAND TIGHTEN THE CONE CLOCKWISE. UNSCREW ANTI-CLOCKWISE FOR REMOVAL.

2. ALTERNATIVE TO THE NOSE CONE INSTALLATION IS TO USE A 1.5 m LONG, CAPPED, RIGID (STRAIGHT) PE PUP FUSED AT FRONT OF THE CARRIER PIPE.

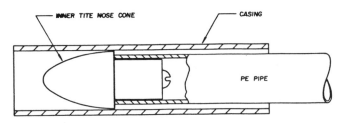

PE PIPE SIZE	INNER TITE MODEL NO.
15.9 mm	M - 2100
26.7 mm	M - 3010
42.2 mm	M - 5010
60.3 mm	M - 5210

Figure 13-16. Nose cone for PE leading end 60.3-mm and smaller PE pipe.

28. Pressure test the services to 770 kPag. Services 42.2 mm and smaller should be given a leak test at 770 kPag using air or nitrogen. Services 60.3 mm and larger should be treated the same as the mains. Soap-test tee caps and all other exposed joints. Wipe all joints clean of soap. Fix any leak found.

Figure 13-17. Main inserted, ready for lateral connection.

Figure 13-18. Main inserted, lateral connection made. Note the sleeve pipe inside the steel casing.

Figure 13-19. New PE service line ready for connection to new main.

29. Tape services and purge air out at the meter stopcock. As each service is tied into the main, test the stopcock core, connect the meter/regulator set, relight the appliances, perform the safety check, and complete the paperwork.

30. Seal the ends of the steel casing pipe using rags, carpet underlay, or similar soft material to provide cushioning for the PE carrier pipe and to stop the ingress of backfill material into the casing. Steel service pipe ends should be taped, and protective sleeves centered over its ends.

31. If the construction drawing calls for anchoring the PE main against lateral movement, provide anchors as shown in Figure 13-14. Wherever the pipe is not encased, i.e., it is exposed, and the backfill material is fillcrete or a similar slurry in which the PE pipe will tend to float unless restrained, install steel U Bars spaced as noted. The details of the steel U Bars are given in Figure 13-14.

Pipe Size		Distance Between U Bars	
mm	in.	m	ft
26.7 to 42.2	3/4 to 1¼	2	6
60.3	2	3	10
≥ 88.9	≥ 3	5	16

32. Check the continuity of the locater wire system to ensure that there are no breaks.
33. Seal off the wall opening around the gas pipe using mortar and/or silicon filler. The unused or retired service-pipe entry should be plugged using a plug or a cap, or the pipe removed and the hole through the wall filled using mortar and/or silicon filler.
34. Where a main valve or a curb cock is installed, place a valve box over the valve and block the valve box independently of the valve. Use precast concrete slab or bricks for blocking.
35. Reinstall, or remove as necessary, valve-location signs in accordance with the company's practices. Note all as-built details prior to back-filling, indicating the locations of all fittings, service tees, underground valves, bends, anchors, etc.

Figure 13-20. Inserted PE pipe emerging at house. Riser bracket installed.

Figure 13-21. Outside riser installed.

Figure 13-22. Riser installed. Bell hole backfilled.

Figure 13-23. A complete service with a multiple meter set.

Figure 13-24. Bell holes being backfilled.

36. Backfill all excavations in accordance with the company specifications (Figure 13-24). Clean and restore site to its original or better condition. Return all surplus materials to the company's stores.
37. Prepare as-builts, material reconciliation, and extra work authorizations. Submit all as-built drawings, inspection reports, etc., to the company.

14
Live Insertion

INTRODUCTION

Live insertion, as the name implies, is a method of gas pipe renewal under live gas conditions. A plastic pipe is inserted into an existing gas main (steel or cast iron) while the latter is still supplying gas to customers. After the new plastic pipe is energized and connected to the customers, the old main is then depressurized and taken out of service. However, it remains in the ground and it contains the new smaller-diameter plastic pipe within it. The advantageous feature of live insertion is that the customers remain connected while the main is replaced. A customer is interrupted only for a short time while his individual service line is reconnected to the new plastic main. Obstructions encountered during the insertion of the PE pipe may require excavating at the obstruction site and cutting out the obstruction. In live insertion, the removal of the obstruction can take place without interrupting the gas supply; in the case of dead insertion, it can cause unforeseen delays in restoring gas supply to the customers.

In live insertion, work can be stopped at any time and at any point of the work without causing a no-gas situation for the remaining customers. This is because all customers remain connected to the old main, which stays alive until such time as the downstream customers are taken off it and reconnected to the PE main. It is this flexibility that makes live insertion more useful, particularly during the heating season or in inclement weather. Live insertion lends itself easily to employing smaller crews who may complete the work at whatever pace the manpower availability will afford.

Although the concept of a sleeve pipe can be adapted to live insertion as well, it is not included. Live insertion described in this chapter is based upon single insertion, i.e., inserting the carrier pipe only, as opposed to double insertion in which the carrier pipe as well as a sleeve pipe are used. In the absence of the sleeve pipe, damage to the carrier pipe during insertion must be carefully considered. Also consider the possibility of freezing if the frost depth in the area goes deeper than the steel casing. Refer to Chapter 6 for a more complete discussion of freezing and scouring problems.

In some countries, patent rights may exist for the live insertion technique or the associated equipment. Check for patents before live inserting.

PLANNING FOR LIVE INSERTION

Since the job-planning work for live and dead insertion is similar, the reader should refer to Chapter 13 for initial job planning. Only the unique variables of the live insertion are discussed here. Generally, inserted PE pipe must operate at a pressure greater than the previous LP steel main. This is to compensate for the decrease in the carrying capacity of the smaller-diameter inserted pipe. Therefore, a higher-pressure gas source must be available at the point where the live insertion project is to commence.

In distribution piping containing excessive existing capacity, raising the system pressure may not be necessary. In that case, live insertion steps would be simpler since you would not need to work from the extremities of the system towards the LP gas source. Insertion can proceed at random anywhere in the system in a piecemeal fashion. This procedure assumes pressure elevation.

The bell-hole sizes for insertion and the receiving end for live insertion are longer than those for dead insertion. A convenient continuous length for live insertion is one city block or about 180 m at a time.

CASING FLOW CALCULATIONS

While sizing for PE carrier pipe can be easily determined using a network analysis program, it is also necessary to calculate the flow capacity of the annular space between the steel casing and the PE pipe, as this annular space supplies low-pressure gas to the customers during the period the PE main is inserted and before the individual services are transferred over to the PE main. (See Tables 14-1 and 14-2.)

Since live insertion work is performed on low-pressure mains operating at 1.72 kPa (¼ psig) or thereabout, a low-pressure flow formula can be used

with reasonable accuracy. Spitzglass' low-pressure formula can be modified
to show that:

$$\Delta P^2 = RLQ^2$$

where $\Delta P^2 = P_1^2 - P_2^2$

P_1 = Absolute inlet pressure kPa

P_2 = Absolute outlet pressure kPa

L = Length of casing pipe in meters

Q = Flow in m³/hr

R = A constant depending on certain conditions

$= 8.265 \times 10^{-3} (D + 3.6 + 0.03\ D^2)\ D^{-6}$

If D, the inside dia. of pipe is in mm

Table 14-1
Determination of R for Annulus

Pipe Size NPS	Steel			PE		
	ID (in.)	(mm)	R	OD (in.)	(mm)	R
12	12.312	312	4.853×10^{-8}	12.75	323	4.08×10^{-8}
10	10.374	263	1.141×10^{-7}	10.75	273	9.54×10^{-8}
8	8.249	209	3.644×10^{-7}	8.625	219	$2.9\ \times 10^{-7}$
6	6.313	160	1.450×10^{-6}	6.625	168	11.28×10^{-7}
4	4.250	107	1.177×10^{-5}	4.5	114	8.66×10^{-6}
3	3.250	82	5.027×10^{-5}	3.5	88	3.35×10^{-5}
2	2.157	54	4.839×10^{-4}	2.375	60	2.83×10^{-4}
1¼	1.380	35	6.028×10^{-3}	1.66	42	2.11×10^{-3}
¾	0.824	20	1.174×10^{-1}	1.05	26	28.88×10^{-3}

Note: 1. R based on Spitzglass low pressure formula.
2. The following conditions were assumed to calculate R for various pipe sizes.
   - Casing pipe, clean, straight and obstruction free.
   - PE pipe suspended co-axially in the casing pipe.
   - P_b-base pressure = 95.29 kPa
   - T_b-base temperature = 10°C
   - P_a-atmospheric pressure = 93.56 kPa
   TF = flowing gas temperature = 10°C
   G = Gas specific gravity = 0.6
   μ = gas viscosity = 1.06×10^{-5} Pa·s

Table 14-2
Determination of R for Annulus

The constant R for the annular space in the casing may be estimated as follows:

$$R \text{ (annulus)} = \frac{R_{ID\text{-}st} \times R_{OD\text{-}PE}}{(\sqrt{R_{OD\text{-}PE}} - \sqrt{R_{ID\text{-}st}})^2} \qquad \text{Hayat's Equation} \quad (5)$$

Size Combination NPS	R annulus
12/6	7.73×10^{-7}
12/4	5.67×10^{-8}
10/6	2.45×10^{-7}
10/4	1.456×10^{-7}
8/6	1.95×10^{-6}
8/4	5.766×10^{-7}
8/3	4.54×10^{-7}
6/4	4.15×10^{-6}
6/3	2.31×10^{-6}
6/2	1.68×10^{-6}
4/2	1.86×10^{-5}
4/1¼	1.37×10^{-5}
3/2	1.50×10^{-4}
3/1¼	7.029×10^{-5}
2/1¼	1.78×10^{-3}
2/¾	6.385×10^{-4}

Example 14-1

Find out the maximum gas flow at the end of the block (180 m), through the annulus of a steel gas main 219-mm OD (8 in.) size in which a 114-mm (4 in.) PE pipe is inserted. Assume initial pressure, 1.72 kPag, and maximum allowable pressure drop, 0.5 kPa. Atmospheric pressure = 93.56 kPa.

$$\therefore P_1 = 93.56 + 1.72 = 95.28 \text{ kPa}$$

$$P_2 = 93.56 + 1.22 = 94.78 \text{ kPa}$$

$$L = 180 \text{ m}$$

$$\therefore \Delta P^2 = P_1^2 - P_2^2 = 95.28^2 - 94.78^2 = 95.03$$

R 8/4 = 5.766 × 10^{-7} (from Table 14-2)

∴ ΔP^2 = R L Q^2

95.03 = 5.766 × 10^{-7} × 180 × Q^2

Q = 956.8 m^3/hr

An efficiency factor (η) to account for the rust, dust, condensate, shavings, thicker wall pipe, and whether it is steel or cast iron should be applied. This may vary between 0.5 to 0.9 depending upon the above variables.

Assuming η = 0.5

Q = 0.5 × 956.8 = 478.4 m^3/hr = 16.89 Mcfh

INSERTION STEPS: STRAIGHT RUN

The following steps are based on the use of polyurethane foam for flow-stopping the low-pressure mains, and on the use of the insertion and receiving fitting designs shown in Figure 14-1. Functionally similar fittings are available from several sources. A patented fitting developed by British Gas and marketed by Steve Vick Limited is shown in Figure 14-2. Minor variations in the procedure will result if Steve Vick's gland boxes are used.

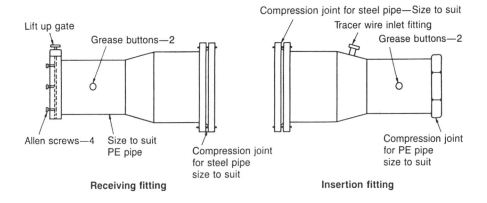

Figure 14-1. Typical insertion and receiving fittings.

Figure 14-2. Gland boxes. (Courtesy of Steve Vick Limited, England.)

After all necessary planning is complete, i.e., approved drawings, man-power, equipment, and materials are available, proceed as follows:

1. Obtain all necessary permits, arrange access to customer's premises. Excavate a bell hole at each service tee, and at both ends of the main, i.e., both at the MP gas source and at the end of the block where the receiving fitting will be installed.
2. Expose steel pipe to allow drilling for a stopper/bag hole at the insertion end and for the stopper/bag hole and the bypass connection at the receiving hole. The steel pipe exposed according to Table 14-3 should provide sufficient working room.

Table 14-3
Recommended Bell Hole Sizes

PE Size	Bell Hole Size (m)			
mm	Service Tee L × W	Insertion L × W	Receiving L × W	Combined Ins./Rec. L × W
42.2 (1¼ in.)	1.5 × 1.0	2.2 × 1.0	2.8 × 1.0	3.4 × 1.0
60.3 (2 in.)	1.5 × 1.0	3.1 × 1.0	2.8 × 1.0	4.3 × 1.0
88.9 (3 in.)	1.5 × 1.0	*4.3 × 1.0	3.3 × 1.0	*5.5 × 1.0
114.3 (4 in.)	1.5 × 1.5	*5.5 × 1.5	3.3 × 1.5	*6.7 × 1.5
168.3 (6 in.)	1.5 × 2.0	*7.9 × 2.0	4.3 × 2.0	*9.1 × 2.0
218.1 (8 in.)	1.5 × 2.0	*10.3 × 2.0	4.3 × 2.0	*11.5 × 2.0

Notes: 1. As the insertion advances, a receiving hole from a previous insertion may become the insertion hole for the next insertion. If that is the case, use the combined insertion/receiving hole figures.

2. Bell hole sizes given in this table are based on an assumed depth of cover of 1.2 m, and on PE tie-in connection by the electrofusion method. For greater cover or for joining by heater plates, bigger bell holes will be necessary due to the requirement to keep the minimum radius of curvature of PE pipe to 25D.

* Pipe pushing in live insertion requires more effort. Except for very short lengths, pushing PE pipe in by hand for sizes 88 mm (3 in.) and larger is impractical. For these sizes, mechanical pipe pushers are necessary. Allow an additional 1.5-m bell hole length if a mechanical pipe pusher is to be used.

3. Install a bypass at the low-pressure source (receiving end). The size of the bypass should be sufficient to carry the expected load. (See Table 14-4.)

4. Stop off the main. Follow the stopper installation and removal sequence given in Figure 14-3. Recommended bag hole sizes are given in Table 14-5.

5. Purge the bypass, then open the bypass to supply gas to the section of the casing to be inserted.

6. Install a stopper at the receiving end (locations 1 and 2). Check the gas pressure.

7. Install another stopper (3) downstream of the last service at the insertion end. Install the insertion fitting.

8. Install a jumper cable before any steel pipe section is cut away. Cut out an appropriate length of steel casing at the receiving end. (Refer to Table 14-6 for steel casing cutout lengths.) Cap off the low-pressure steel main that is to stay in service. Install the receiving fitting with its ends closed off. Remove stopper (1) and plug hole with a steel plug. Remove stopper (2) and install a sight glass at (2).

9. Mark PE pipe with marker every 5 m. This step is not necessary if a mechanical counter is fitted on the pipe pushing machine.

Table 14-4
*Flow Through Low Pressure Bypass – m³/hr.

Bypass Length (m)	Size						
	26 mm	42 mm	60 mm	88 mm	114 mm	168 mm	219 mm
5	5.14	22.70	80.12	248.57	513.70	1,463.58	2,919.52
10	3.64	16.05	56.65	175.76	363.24	1,034.91	2,064.41
15	2.97	13.11	46.26	143.51	296.50	845.00	1,685.59
20	2.57	11.35	40.06	124.28	256.85	731.79	1,459.76
25	2.30	10.15	35.83	111.16	229.74	654.53	1,305.65
30	2.10	9.27	32.71	101.48	209.72	597.50	1,191.89
35	1.94	8.58	30.28	93.95	194.16	553.18	1,103.48
40	1.82	8.03	28.33	87.88	181.62	517.45	1,032.21
45	1.71	7.57	26.71	82.86	171.23	487.86	973.17
50	1.63	7.18	25.34	78.60	162.45	462.82	923.23
60	1.48	6.55	23.13	71.76	148.29	422.50	842.79
70	1.37	6.07	21.41	66.43	137.29	391.16	780.27
80	1.29	5.67	20.03	62.14	128.43	365.90	729.88
90	1.21	5.35	18.88	58.59	121.08	344.97	688.14
100	1.15	5.08	17.91	55.58	114.87	327.27	652.82
110	1.10	4.84	17.08	53.00	109.52	312.04	622.44
120	1.05	4.63	16.35	50.74	104.86	298.75	595.94
130	1.01	4.45	15.71	48.75	100.75	287.03	572.57
140	0.97	4.29	15.14	46.98	97.08	276.59	551.74
150	0.94	4.14	14.63	45.38	93.79	267.21	533.03
160	0.91	4.01	14.16	43.94	90.81	258.73	516.10

* Basis—Gas temperature 10°C, ambient pressure 95.29 kPa, gas gravity 0.6, pressure drop over the length of bypass 0.08 kPa from 1.82 kPa to 1.74 kPa, using Spitzglass low pressure formula. The flow is based on straight lengths of steel pipe; allow for bends and valves, etc.

10. Thread (TWU 14) epoxy-coated copper tracer wire through the insertion fitting and attach it to the PE pipe cap. Push in pretested PE pipe past the insertion fitting about 0.2 m. Since the tracer wire is attached to the PE pipe, it will travel with it. If the casing pipe is steel and of welded construction, the tracer wire may not be used as the steel casing itself can be used for locating.

11. Remove stopper (3) and plug off hole using a steel plug. After the stopper is removed, some gas may start to escape through the insertion fitting. Proper tightening of glands supplemented by the injection of grease through the grease nipple will minimize the gas escape.

12. Ensure that men working in the trench are using breathing apparatus, as they will be working with live gas. Ensure that fire extinguishers are available at the bell holes.

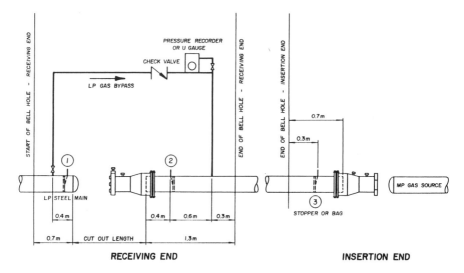

Figure 14-3. Bypass and stopper installation.

Table 14-5
Bag Hole Sizes for Main

Steel Main Size		Bag Hole Size
(mm)	(in.)	NPS
60	2	3/4
88	3	1 1/4
114	4	1 1/2
168 to 323	6 to 12	2

13. Push PE pipe ahead towards the receiving end. 60-mm (2-in.) and smaller PE pipe can be pushed in by hand easily. 88-mm (3-in.) and larger sizes require mechanical pipe pushers. (See Figure 14-4.) Inject grease at the insertion fitting to minimize gas escape through the fitting. If the pipe hangs up during insertion, rotate it 180° and repush.

This action should generally free the pipe end for further travel. Failure to free the pipe means there is an obstruction, which must be removed. The marking made on the PE pipe will help determine the end of the inserted pipe. Excavate at the obstruction, pull the PE pipe back 2 m, install another LP gas bypass at the obstruction, and install stoppers. Cut out the obstruction and piece through the steel main. Remove stoppers and the bypass. Continue with the insertion.

14. Push PE pipe until it moves past the sight glass and into the receiving fitting up to the gate.

Table 14-6
Steel Casing Cutout Lengths

PE Size	Steel Pipe Cut Length (m)		
(mm)	Insertion	Receiving	Service Tee
42.2	1.4	1.4	0.9
60.3	2.3	1.4	0.9
88.9	3.5	1.9	0.9
114.3	4.7	1.9	0.9
168.3	7.1	2.9	0.9
218.1	9.5	2.9	0.9

Figure 14-4. Alvic mini-mechanical pipe pusher. (Courtesy of Steve Vick Limited, England.)

15. Open the receiving fitting door and push the PE pipe through it, about 1 m past the fitting. Allow sufficient length of PE pipe at the insertion end to enable connection to the MP gas source. Grease as required in the insertion and the receiving fittings to minimize gas leakage through the fittings.

16. Remove sight glass at 2 (Figure 14-5) and foam in place, directing the foaming nozzle towards the insertion end. Ensure that the quantity of foam is such that it forms a foam plug of 0.2 to 0.3 mm long. Check the pressure recorder. Do not overfoam. Overfoaming may block the bypass.

17. Remove the receiving fitting, and foam the end of the steel casing. This provides double sealing and ensures that no gas will seep through. Similarly, foam through stopper hole 3 (Figure 14-5), re-move insertion fitting, and foam casing's end off.

18. Using nitrogen or bottled natural gas, give the PE pipe a final leak test of 15 minutes at the appropriate test pressure. This final test is given to ensure that the insertion did not cause unacceptable injury to the plastic pipe. Air should not be used for this test as an air leak may create explosive mixtures in the gas stream, and may also put out the pilot lights in the customers' appliances. If the final pressure test fails, reschedule the work using the dead insertion method.

19. Connect the PE pipe to the MP gas source at the insertion end. This connection may be made by the butt-fusion or the electrofusion method.

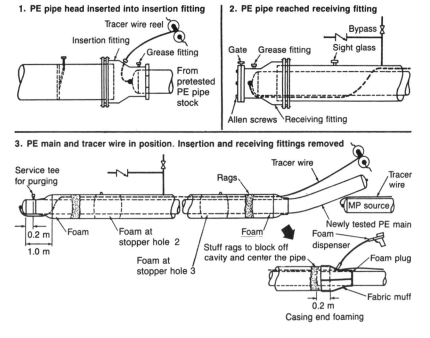

Figure 14-5. PE main travel during insertion.

20. If nitrogen was used for testing, install a service tee or a purge tee at the receiving end and purge the nitrogen out.

21. About 0.6 m (2 ft) upstream of the last service tee, drill a 12-mm (1/2-in.) hole in the casing for the foam injection. Turn off gas to the last customer. Foam the main upstream of the last service to isolate LP gas supply. Refer to Figure 14-6.

22. Install a jumper cable before any steel pipe section is cut away. Cut off the service pipe 1 m away from the service tee, or if the service line has a curb valve, up to and including the curb valve. Remove the service tee and cut a cylindrical portion of steel casing, 0.9 m, centered over the service tee. Since the casing now contains the PE pipe, cutting out the cylindrical portion can only be done in steps. Using an axial pipe cutter, first make two axial cuts at the 3 and 9 o'clock positions, then use a circular pipe cutter to make two cuts at the top, circumferentially removing the top part of the casing. Lift the PE pipe from the bottom, block it by wood or rags, then saw off the bottom portion of the casing. Ensure that the depth of the cutter wheel is such that it cuts only the steel pipe. Check the PE pipe's surface for scratches. Figures 14-6 and 14-7 show the pipe cutting operation for steel casing. In the case of a cast iron casing, no axial cuts are needed.

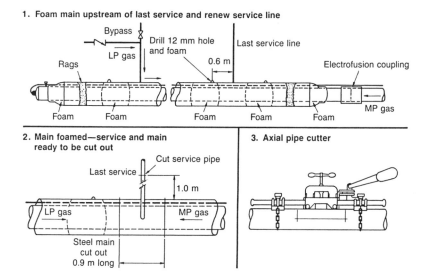

Figure 14-6. Renewing services.

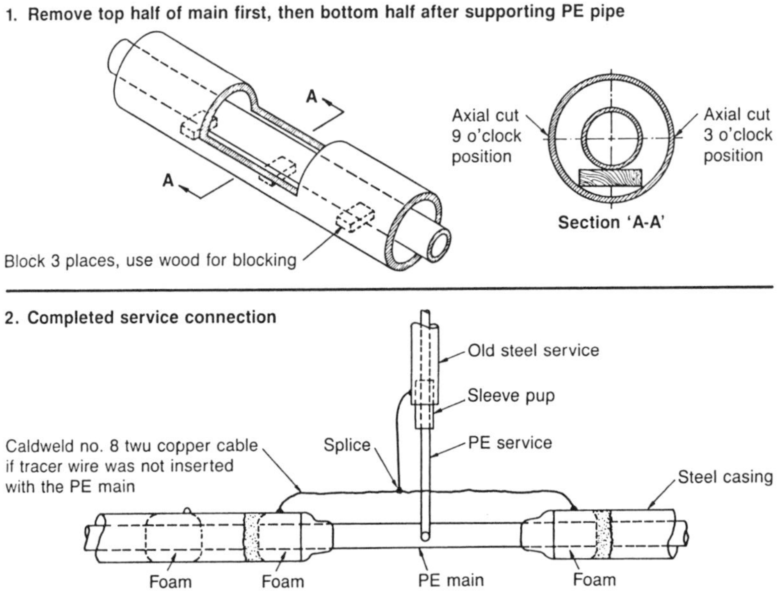

1. Remove top half of main first, then bottom half after supporting PE pipe

Axial cut
9 o'clock
position

Axial cut
3 o'clock
position

Section 'A-A'

Block 3 places, use wood for blocking

2. Completed service connection

Old steel service

Sleeve pup

Caldweld no. 8 twu copper cable
if tracer wire was not inserted
with the PE main

Splice

PE service

Steel casing

Foam   Foam   PE main   Foam

Figure 14-7. Service tee connection to PE main.

Simply make the circumferential cuts first, then smash the cylindrical portion out by crushing it with a pipe squeezer. Remove debris from the trench.

23. Lift up the exposed PE pipe and foam both ends. The use of foam obviates the need for the provision of sleeve pups at the casing's end.

24. If a tracer wire was not used, cadweld No. 8 TWU copper cable across the cut to bridge the discontinuity caused by the removal of a section of the steel pipe. Refer to Figure 14-7.

25. Perform the service insertion by the dead insertion method. Dead insertion was described in Chapter 13. Test the service line and connect it to the PE main. Connect the service-line tracer wire to the main-line tracer wire.

26. Anchoring the PE pipe is necessary due to the differential expansion and contraction of the PE and steel pipes. Anchor the PE pipe every 30 m. Anchoring may be achieved by the installation of external anchors, such as a Dresser-style 180 clamp, or by a block of concrete or an adequate backfill. Adequate natural soil anchoring is attained if the service tee bell holes are less than 15 m apart, or if at least one bell hole in every 30 m of pipe run is backfilled using a low-strength slurry, such as fillcrete or any stronger backfill material. If inserted

plastic pipe is not anchored, any lateral connection, such as a service tee, may undergo severe shear stress and may be sheared off. This can be explained by the following example:

Coefficient of linear expansion of plastic = 0.15 mm/m/°C

Coefficient of linear expansion of steel = 0.012 mm/m/°C

Therefore, the differential expansion = 0.15 − 0.012 = 0.138 mm/m/°C

For a 30 m length, with 20°C temperature change between the summer and winter pipe temperature, the differential change in plastic pipe length:

$$0.138 \times 30 \times 20 = 82.8 \text{ mm}$$

Figure 13-14 shows the application of an anchor clamp.

27. Similarly, other service lines may be renewed one at a time by moving towards the source of LP gas. Conditions permitting (that is, available manpower, acceptable weather, convenient time of day, and acceptance by customers), more than one customer at a time may be shut off to enable service line work to proceed simultaneously rather than by one service at a time. In this case, the intermediate foaming holes near the service tees are not required.

28. Abandon inactive services by cutting-off service lines near the main; remove curb stops if they exist and plug service pipe with steel bungs, one at each end of the cut service pipe.

29. Since the service line renewal work must proceed in a sequential order, one at a time or in batches, while proceeding towards the source of the LP gas, inaccessible premises should have their service lines renewed and terminated outside with a riser, ready for the meter installation and connection to the inside piping when access is available.

30. Once all services are renewed and connected to the PE main, the bypass can be removed. Live insertion work in this block is now complete. Proceed with backfilling, clean up, and as-built preparation.

INSERTION AT LATERALS

Live lateral insertions can be made the same way as the longitudinal runs. MP gas has to be brought up to the tee junction. This can be done by situating the receiving hole at the tee junction and running the PE main just past the tee junction.

However, before the steel tee can be removed, an additional LP bypass to the lateral must first be fabricated and a stopper installed in the lateral to isolate the tee. Finish the live insertion on the main run in the usual way, while supplying LP gas to the lateral through the lateral bypass. Insert the PE main in lateral, test the PE lateral, make the connection to the PE main run by installing a suitable-size PE tee. Once MP gas is available in the lateral, the services in the lateral can be switched to MP main starting from the farthest end of the lateral. Note that in the case of the lateral, both the LP gas and the MP gas flow in the same direction and not in opposite directions. All other steps are the same as for the insertion in straight runs. Figure 14-8 refers to insertion at laterals. Install anchors at all main laterals or fill bell holes with fillcrete.

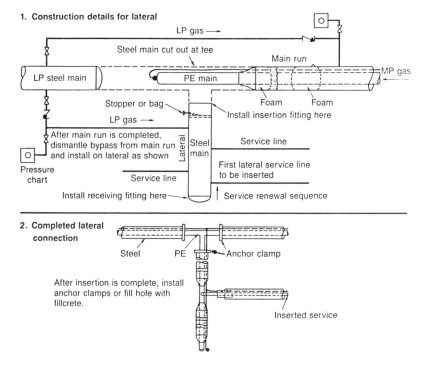

Figure 14-8. Insertion at laterals.

FOAMING

Polyurethane foams can be used to stop the flow of gas in low-pressure (1.72 kPa) and medium-pressure (210 kPa) gas pipelines. Development now being done in the industry may make it possible to use foams for flow stopping at higher pressures. The foaming consists of two-part chemicals which are stored separately and are brought together under controlled conditions to a mixing chamber, where they are mixed together in the correct proportion and injected via a nozzle into the gas pipeline. Upon activation, the chemicals may expand to over 50 times their previous volume. By varying the composition, the physical properties of the foam can be changed from very soft and rubber-like to hard and brittle. The chemicals rapidly rise and, within a few seconds, create a blockage within the pipe. The chemicals used in the polyurethane foams are quite tolerant to the presence of gas contaminants such as water, glycol, and heavier hydrocarbons.

One-part chemical foams, which depend on moisture for activation, are also available. These are slow to react and cure, especially in enclosed voids. Ureaformaldehyde foam is also known to have been used in the gas mains in the past, but is no longer used. The description given in this chapter pertains to polyurethane foams.

Portable Foaming Equipment

There is a choice of equipment and methods. Some specialized equipment can inject foam under pressure without exposing the operator to live gas. Packages are available in small kits that are specially formulated to do a single foaming operation for a particular pipe size, using a foaming cartridge and a hand-operated cartridge gun (see Figures 14-9 and 14-10). Similarly, in the mid-size range, a 16-kg foam kit, complete with nozzle and hoses, is available. This kit is sufficient to do a block of main and service work.

Bulk Foaming Procedure

Before trying actual foaming, ensure that the foaming equipment is working and that the main is ready for foaming. Then follow the steps given below.

1. To ensure that the foam is suitable, eject a small amount of mixture from the nozzle and observe the curing. The test foam should set in 3-5 seconds in warm weather, and attain a cream to light-brown color. If it does not happen, check the equipment operation.

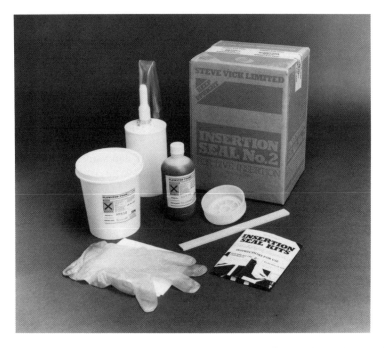

Figure 14-9. Prepacked foaming kits. (Courtesy of Steve Vick Limited, England.)

2. When everything is ready, remove plug from main, inject foam, and cover the tap-hole until the foam sets. Direct foam towards the bottom on both side of the plastic pipe.
3. If flow is not stopped completely, make a hole in the foam and inject additional foam underneath the plastic pipe. In the unlikely event that the gas leakage persists, drill another hole upstream in the casing and foam again. If necessary a third foam plug can be created in between the previous two plugs. This will be a much denser plug.
4. Foaming close to a bypass or a valve requires careful attention so that the foam does not interfere with their operation. When foaming close to these, direct the foam jet away from these appurtenances.
5. Foaming at the casing end can be accomplished using a fabric muff as shown in Figure 14-5. Stuff rags at casing ends before foaming in order to center PE pipe and provide an enclosed area. The use of a muff will completely contain the foam during curing.

   Containing foam in a closed area makes denser foam, which resists greater pressure. The fabric muff can be removed and used again.

Figure 14-10. Prepacked foaming kits. (Courtesy of Insta-Foam Products Inc., USA.)

6. Polyurethane foams, after the initial curing, can shrink slightly during the first 24 hours. All foam plugs holding pressure at pipe ends should be reinforced with adhesive tape, and mechanically restrained (with a reducing collar or a similar fitting) if left unattended.
7. Clean and store equipment safely after use.

Chemical Components of Polyurethane Foams

Polyurethane foams are generally composed of two reactive mixtures—an isocyanate component (A component) and a blended polyol resin (B component), supplied either as drum sets or in bulk to large-volume users with tankage facilities. In use, the two components are metered in the proper ratio of A and B through proportioning pumps to a mixing head or spray gun, where they are intimately mixed and dispensed to produce a finished foam. Although now relatively rare, some foam systems may include a third liquid component, usually a catalyst blend, which, for any of several reasons, is packaged to be fed as a third stream at the applications site. The generic chemical compositions of polyurethane foam systems are listed in Table 14-7 with indicated potential hazards.

Table 14-7
Polyurethane Foams

Component	Chemical Composition	Skin Irritant*	Potential Sensitizer (Respiratory or Skin)
A Isocyanate	Monomeric, polymeric, or propolymer	Yes	Yes
B Resin Blend	Polyol resin	No	No
	Amines and/or metallic salt catalysts	Yes	Yes (some)
	Chlorofluorocarbon blowing agents	No	No
	Silicone surfactants	No	No
C Catalyst Blend	Amines	Yes	Yes (some)
	Water	No	No

* Note: All components, either in liquid or vapor form, can cause injury to the eyes. During all spray applications of polyurethane foam, positive pressure air respirators must be worn.

Solvents are used in cleanup operations and in flushing polyurethane-foam dispensing equipment. These solvents may be toxic, flammable, or irritants, and the solvent supplier should be consulted regarding the safe handling of any solvents used.

REFERENCES

1. The Society of the Plastic Industry, Inc., New York, SPI Bulletin U118-R.
2. Insta Foam Products—Calgary, Alberta, Canada.
3. Steve Vick, Limited—Somerset, England.

Appendix

Table A-1
Commonly Used Steel Pipe for Distribution

Nominal Diameter (mm)	Wall Thickness (mm)	Mass/ Meter (kg/m)	Grade (MPa)	Mill Test (MPa)	Type	Nominal Diameter (in.)	Wall Thickness (in.)
15 tubing	0.89	0.38	172	10.30	Seamless	1/2	0.035
26 STD (40)(RR)(HWY)	2.87	1.68	172	4.83	CW/ERW	3/4	0.113
42 STD (40)(RR)(HWY)	3.56	3.38	207	6.90	CW/ERW	1 1/4	0.140
60 Line pipe (HWY)	2.77	3.93	290	15.93	ERW	2	0.109
60 STD (40)(RR)	3.91	5.44	290	20.70	ERW	2	0.154
88 Line pipe (HWY)	3.18	6.72	290	12.40	ERW	3	0.125
88 (RR)	4.78	9.91	290	18.69	ERW	3	0.188
88 STD (40)	5.49	11.29	290	20.70	ERW	3	0.216
114 Line pipe (HWY)	3.18	8.70	290	9.66	ERW	4	0.125
114 (RR)	4.78	12.90	290	14.58	ERW	4	0.188
114 STD (40)	6.02	16.07	290	18.27	ERW	4	0.237
168 Line pipe	3.96	16.07	290	10.21	ERW	6	0.156
168 (RR)(HWY)	4.78	19.24	290	12.35	ERW	6	0.188
168 STD (40)	7.11	28.26	290	18.35	ERW	6	0.280
219 Line pipe (RR)(HWY)	4.78	25.23	290	9.45	ERW	8	0.188
219 (20)	6.35	33.31	290	12.61	ERW	8	0.250
219 STD (40)	8.18	42.53	290	16.21	ERW	8	0.322
273 Line pipe (RR)(HWY)	4.78	31.59	290	8.62	ERW	10	0.188
273	7.09	46.47	290	12.75	ERW	10	0.279
273 STD (40)	9.27	60.29	290	16.70	ERW	10	0.365
323 Line pipe (RR)(HWY)	5.56	43.66	290	8.49	ERW	12	0.219
323	7.92	61.74	290	12.07	ERW	12	0.312
323 STD	9.52	73.82	290	14.51	ERW	12	0.375
323 (40)	10.31	79.72	290	15.86	ERW	12	0.406
406 Line pipe (RR)(HWY)	5.56	54.98	290	6.76	ERW	16	0.219
406	7.92	77.86	290	9.60	ERW	16	0.312
406 STD	9.52	93.21	290	11.51	ERW	16	0.375
406 (40)	12.70	123.29	290	15.37	ERW	16	0.500
508 Line pipe (HWY)	6.35	78.54	290	6.56	ERW	20	0.250
508 (RR)	7.14	88.15	290	7.31	ERW	20	0.281

191

Table A-1 Continued

508	7.92	97.71	290	8.15	ERW	20	0.312
508 STD	9.52	117.07	290	9.80	ERW	20	0.375

Note:

STD-Standard Wall   40-sch 40   RR-Suitable for railroad crossings

HWY-Suitable for highway crossing   Line Pipe-commonly used

Table A-2
Plastic Pipe and Tubing Dimensions

Nominal Size (in.)	Nominal Size (mm)	Minimum Wall Thickness (mm)	Average Outside Diameter (in.)	Minimum Wall Thickness (in.)
		SDR 11		
3/4	26.7	2.42	1.050	0.095
1	33.4	3.04	1.315	0.119
1 1/4	42.2	3.82	1.660	0.151
1 1/4 (SDR 10)	42.2	4.22	1.660	0.166
1 1/2	48.3	4.38	1.900	0.173
2	60.3	5.48	2.375	0.216
2 1/2	73.0	6.62	2.875	0.261
3	88.9	8.08	3.500	0.318
4	114.3	10.38	4.500	0.409
6	168.3	15.28	6.625	0.602
		SDR 8.8		
3/4	26.7	3.03	1.050	0.119
1	33.4	3.79	1.315	0.149
1 1/4	42.2	4.80	1.660	0.189
1 1/2	48.3	5.48	1.900	0.216
2	60.3	6.86	2.375	0.270
2 1/2	73.3	8.30	2.875	0.327
4	88.9	10.10	3.500	0.398
4	114.3	12.98	4.500	0.511
6	168.3	19.12	6.625	0.753
		SDR 13.5		
2	60.3	5.00	2.375	0.197
4	114.3	8.46	4.500	0.333
6	168.3	12.46	6.625	0.491
		SDR 21		
6	168.3	8.03	6.625	0.316
		TUBING		
1/2 CTS	15.9	1.58	0.625	0.062
1/2 CTS	15.9	2.28	0.625	0.090
1 CTS	28.6	1.58	1.125	0.062
1 CTS	28.6	2.50	1.125	0.099

SDR = Standard Dimension Ratio = OD Divided by Wall Thickness

Table A-3
Steel Line Pipe and Component Size Nomenclature

Pipe size OD, mm	Nominal size of matching fitting, flange, or valve			Pipe size OD, mm	Nominal size of matching fitting, flange, or valve		
10.3	NPS	1/8	DN 6	762	NPS 30	DN	750
13.7	NPS	1/4	DN 8	813	NPS 32	DN	800
17.1	NPS	3/8	DN 10	864	NPS 34	DN	850
21.3	NPS	1/2	DN 15	914	NPS 36	DN	900
26.7	NPS	3/4	DN 20	965	NPS 38	DN	950
33.4	NPS	1	DN 25	1016	NPS 40	DN	1000
42.2	NPS	1 1/4	DN 32	1067	NPS 42	DN	1050
48.3	NPS	1 1/2	DN 40	1118	NPS 44	DN	1100
60.3	NPS	2	DN 50	1168	NPS 46	DN	1150
73.0	NPS	2 1/2	DN 65	1219	NPS 48	DN	1200
88.9	NPS	3	DN 80	1270	NPS 50	DN	1250
101.6	NPS	3 1/2	DN 90	1321	NPS 52	DN	1300
114.3	NPS	4	DN 100	1372	NPS 54	DN	1350
141.3	NPS	5	DN 125	1422	NPS 56	DN	1400
168.3	NPS	6	DN 150	1473	NPS 58	DN	1450
219.1	NPS	8	DN 200	1524	NPS 60	DN	1500
273.1	NPS	10	DN 250	1575	NPS 62	DN	1550
323.9	NPS	12	DN 300	1626	NPS 64	DN	1600
355.6	NPS	14	DN 350	1676	NPS 66	DN	1650
406.4	NPS	16	DN 400	1727	NPS 68	DN	1700
457.0	NPS	18	DN 450	1778	NPS 70	DN	1750
508.0	NPS	20	DN 500	1829	NPS 72	DN	1800
559.0	NPS	22	DN 550	1880	NPS 74	DN	1850
610.0	NPS	24	DN 600	1930	NPS 76	DN	1900
660.0	NPS	26	DN 650	1981	NPS 78	DN	1950
711.0	NPS	28	DN 700	2032	NPS 80	DN	2000

Notes:
1. 'NPS' means 'nominal pipe size' and the NPS system of nominal size designation is contained in standards prepared by the American Society of Mechanical Engineers. The NPS size is dimensionless and the numerical portion of the designation is identical to the numerical portion of the previously used inch nominal size designation.

2. The 'DN' means 'diameter nominal' and the DN system of nominal size designation is contained in standards prepared by the International Organization for Standardization (ISO).

3. The DN nominal sizes shown in this table have generally been extracted from various ISO standards, but in some cases, have been assigned arbitrarily. Caution should be exercised in the use of this table, since in many cases, the DN nominal size shown is identical to that used in ISO standards to designate components for pipe having a specified outside diameter that differs slightly from the OD size shown.

Reference: Steel Line Pipe. CSA Standard CAN 3-Z 245.1-M86.

Table A-4
Copper Tube Sizes

OD (mm)	Nominal Size (mm)	English (CTS) (in.)	OD (in.)
9.5	NTS 1/4	1/4	3/8
12.7	NTS 3/8	3/8	1/2
15.9	NTS 1/2	1/2	5/8
28.6	NTS 1	1	1 1/8

Table A-5
Stainless Steel Tube Sizes

OD (mm)	Nominal Size Metric (mm)	English (SST) (in.)	OD (in.)
6.4	6.4 OD-SST	1/4	3/8
9.5	9.5 OD-SST	3/8	3/8
12.7	12.7 OD-SST	1/2	1/2

Table A-6
Valves and Fittings Pressure Ratings

Metric Designation	Max. Pressure Rating (kPa)*	ANSI, WOG, or CI Class	Max. Pressure Rating (psig)*
Class 125	1,030	ANSI 125	150
125 CI	1,200	CI 125	175
200 WOG	1,200	WOG 175	175
Class 150 MI	2,070	MI 150	300
	Up to 65°C	Up to 150°F	
Class 150 (PN 20)	1,965	ANSI 150	285
Class 300 MI	NPS 1/4–1	MI 300	1/4 in.–1 in.
	13,800		2,000
	Up to 65°C		Up to 150°F
	NPS 1 1/4–2		1 1/4–2
	10,340		1,500
	Up to 65°C		Up to 150°F
250 CI	2,760	CI 250	400
400 WOG	2,760	WOG 400	400
Class 300 (PN 50)	5,100	ANSI 300	740
2000 FS	13,800	FS 2000	2,000
3000 FS	20,700	FS 3000	3,000
Class 600 (PN 100)	10,200	ANSI 600	1,480

* − 30°C to + 40°C
Note: ISO designation is by PN (pressure nominal).

Table A-7
Prefixes

Magnitude	Multiplying Factor	Prefix	Symbol
Quintillion	1 000 000 000 000 000 000 = 10^{18}	exa	E
Quadrillion	1 000 000 000 000 000 = 10^{15}	peta	P
Trillion	1 000 000 000 000 = 10^{12}	tera	T
Billion	1 000 000 000 = 10^{9}	giga	G
Million	1 000 000 = 10^{6}	mega	M
Thousand	1 000 = 10^{3}	kilo	k
Hundred	100 = 10^{2}	hecto	h
Ten	10 = 10^{1}	deca	da
Tenth	0.1 = 10^{-1}	deci	d
Hundredth	0.01 = 10^{-2}	centi	c
Thousand	0.001 = 10^{-3}	milli	m
Millionth	0.000 001 = 10^{-6}	micro	u
Billionth	0.000 000 001 = 10^{-9}	nano	n
Trillionth	0.000 000 000 001 = 10^{-12}	pico	p
Quadrillionth	0.000 000 000 000 001 = 10^{-15}	femto	f
Quintillionth	0.000 000 000 000 000 001 = 10^{-18}	atto	a

Table A-8
Common Gas Industry Conversion Factors

Quantity	Metric Unit	Symbol	U.S. Equivalent
Length	meter	m	3.281 ft
Area	square meter	m^2	10.763 ft^2
Volume	cubic meter	m^3	35.32 ft^3
Temperature	degree Celsius	°C	5/9 (°F − 32)
Degree day deficiency	Celcius	DDDC	DDDF × 5/9
Pressure	kilopascal	kPa	0.145 psi (lbf/in.^{2})
Stress	megapascal	MPa	145 psi (lbf/in.^{2})
Force	newton	N	0.2248 lbf
Torque	newton meter	N·m	0.7375 ft lbf
Liquid measure	liter	L	0.264 U.S. gallons
Heat energy	megajoule	MJ	948.213 Btu
	gigajoule	GJ	9.482 therm
Heat Rate/Power		MJ/h	0.3725 hp
		kW	1.341 hp
Mass	tonne	t	2205 lb

Table A-9
SI Units

Unit Type	Quantity	Unit	Symbol	Expression in Other Units
Base Units	Length	meter	m	
	Mass	kilogram	kg	
	Time	second	s	
	Electric current	ampere	A	
	Thermodynamic temperature	kelvin	K	
	Amount of substance	mole	mol	
	Luminous intensity	candela	cd	
Supplementary Units	Plane angle	radian	rad	
	Solid angle	steradian	sr	
	Frequency	hertz	Hz	s^{-1}
	Force	newton	N	$kg{\cdot}m/s^2$
	Pressure, stress	pascal	Pa	N/m^2
	energy, work quantity of heat	joule	J	$N{\cdot}m$
	Power	watt	W	J/s
	Electric charge	coulomb	C	$A{\cdot}s$
Derived Units with Special Names	Electric potential Potential difference Electromotive force	volt	V	W/A
	Electric resistance	ohm	Ω	V/A
	Electric conductance	siemens	S	A/V
	Electric capacitance	farad	F	C/V
	Magnetic flux	weber	Wb	$V{\cdot}s$
	Inductance	henry	H	Wb/A
	Magnetic flux density	tesla	T	Wb/m^2
	Luminous flux	lumen	lm	$cd{\cdot}sr$
	Illuminance	lux	lx	lm/m^2
	Activity of radionuclides	becquerel	Bq	s^{-1}
	Absorbed dose of ionizing radiation	gray	Gy	J/kg
	Celsius temperature	degree Celsius	°C	K*

* On the temperature scale 0°C = 273.15 K
For temperature intervals 1°C = 1 K

Table A-10
Examples of SI Derived Units without Special Names

Quantity	Unit Name	Unit Symbols Typical Form	Unit Symbols In Base Units
area	square meter	m^2	m^2
volume	cubic meter	m^3	m^3

Table A-10 Continued
Examples of SI Derived Units without Special Names

Quantity	Unit Name	Unit Symbols Typical Form	In Base Units
speed—linear	meter per second	m/s	$m \cdot s^{-1}$
—angular	radian per second	rad/s	$rad \cdot s^{-1}$
acceleration—linear	meter per second squared	m/s^2	$m \cdot s^{-2}$
—angular	radian per second squared	rad/s^2	$rad \cdot s^{-2}$
wave number	reciprocal meter	m^{-1}	m^{-1}
density, mass density	kilogram per cubic meter	kg/m^3	$kg \cdot m^{-3}$
concentration (of amount of substance)	mole per cubic meter	mol/m^3	$mol \cdot m^{-3}$
specific volume	cubic meter per kilogram	m^3/kg	$m^3 \cdot kg^{-1}$
luminance	candela per square meter	cd/m^2	$cd \cdot m^{-2}$
dynamic viscosity	pascal second	Pa·s	$m^{-1} \cdot kgs^{-1}$
kinematic viscosity	square meter per second	m^2/s	$m^2 \cdot s^{-1}$
moment of force	newton meter	N·m	$m^2 \cdot kg \cdot s^{-2}$
surface tension	newton per meter	N/m	$kg \cdot s^{-2}$
heat flux density, irradiance	watt per square meter	W/m^2	$kg \cdot s^{-3}$
heat capacity, entropy	joule per kelvin	J/K	$m^2 \cdot kg \cdot s^{-2} \cdot K^{-1}$
specific heat capacity, specific entropy	joule per kilogram kelvin	$J/(kg \cdot K)$	$m^2 \cdot s^{-2} \cdot K^{-1}$
specific energy	joule per kilogram	J/kg	$m^2 \cdot s^{-2}$
thermal conductivity	watt per meter kelvin	$W/m(m \cdot K)$	$m \cdot kg \cdot s^{-3} \cdot K^{-1}$
energy density	joule per cubic meter	J/m^3	$m^{-1} \cdot kg \cdot s^{-2}$
electric field strength	volt per meter	V/m	$m \cdot kg \cdot s^{-3} \cdot A^{-1}$
electric charge density	coulomb per cubic meter	C/m^3	$m^{-3} \cdot s \cdot A$
surface density of charge, flux density	coulomb per square meter	C/m^2	$m^{-2} \cdot s \cdot A$
permittivity	farad per meter	F/m	$m^{-3} \cdot kg^{-1} \cdot s^4 \cdot A^2$
current density	ampere per square meter	A/m^2	$A \cdot m^{-2}$
magnetic field strength	ampere per meter	A/m	$A \cdot m^{-1}$
permeability	henry per meter	H/m	$m \cdot kg \cdot s^{-2} \cdot A^{-2}$
molar energy	joule per mole	J/mol	$m^2 \cdot kg \cdot s^{-2} \cdot mol^{-1}$
molar entropy molar heat capacity	joule per mole kelvin	$J/(mol \cdot K)$	$m^2 \cdot kg \cdot s^{-2} \cdot K^{-1} \cdot mol^{-1}$
radiant intensity	watt per steradian	W/sr	$m^2 \cdot kg \cdot s^{-3} \cdot sr^{-1}$
radiance	watt per square meter steradian	$W/(m^2 \cdot sr)$	$kg \cdot s^{-3} \cdot sr^{-1}$

Table A-10 Continued
Examples of SI Derived Units without Special Names

Quantity	Unit Name	Typical Form	Unit Symbols In Base Units
exposure	coulomb per kilogram	C/kg	$A \cdot s \cdot kg^{-1}$
absorbed dose rate	gray per second	Gy/s	$m^2 \cdot s^{-3}$

Table A-11
Permissible Non-SI Units

Condition of Use	Unit	Symbol	Value SI Units
	minute	min	1 min = 60 s
	hour	h	1 h = 3600 s
Permissible	day	d	1 d = 86,400 s
Universally	degree (of arc)	°	$1° = (\pi/180)$ rad
with SI	minute (of arc)	′	$1′ = (\pi/10,800)$ rad
	second (of arc)	″	$1″ = (\pi/648,000)$ rad
	litre	L	$1 L = 1 dm^3$
	tonne	t	$1 t = 10^3$ kg
	hectare	ha	$1 ha = 10^4 m^2$
Permissible in specialized fields	electronvolt	eV	1 eV = 0.160 218 aJ
	unit of atomic mass	u	$1 u = 1.660\ 54 \times 10^{-27}$ kg
	nautical mile	M	1 nautical mile = 1852 m
	knot	kn	1 nautical mile per hour =
Permissible for a limited time			(1852/3600) m/s
	hectare	ha	$1 ha = 10^4 m^2$
	millibar	mbar	1 mbar = 100 Pa
	standard atmosphere	atm	1 atm = 101.325 kPa

Note: For a full text, refer to Canadian Metric Practice Guide CAN/CSA Z234.1

Table A-12
Engineering Conversion Factors

U.S.	Sub-division	Area SI	SI	U.S.
1 sq. in.		$= 6.4516 cm^2$	$1 cm^2$	$= 0.1550 in.^2$
1 sq. ft	$1 ft^2 = 144 in.^2$	$= 9.29 dm^2$	$1 m^2$	$= 10.764 ft^2$

	Dynamic Viscosity			
		Pa · S	Poise	lb/ft s
$1 Ns/m^2 =$	1 da Poise	1	10	0.67197
1 Poise		0.1	1	0.067197
1 lb/ft/s		1.48816	14.8816	1

Table A-12 Continued
Engineering Conversion Factors

Energy—Work

	J	kWh	kcal	Btu
1 J = 1 W·s = 1 N·m	1	2.7778×10^{-7}	2.3901×10^{-4}	9.478178×10^{-4}
1 kWh	3.6×10^6	1	860.421	3.41214×10^3
1 kcal*	4.184×10^3	1.1622×10^{-3}	1	3.96567
1 Btu**	1.055056×10^3	2.93071×10^{-4}	0.252164	1

Flow

U.S.	Sub-division	SI	SI	U.S.
1 cu. ft/min	1 cfm	= 1.699 m³/hr	1 m³/hr	= 0.5886 cfm
1 st/day	1 st/d	= 37.80 kg/hr	1 kg/hr	= 0.02646 st/d
1 lb mol/hr	1 lb mol/hr	= 0.4536 k mol/hr	1 k mol/hr	= 2.2046 lb mol/hr
1 scfm at 60°F, 1 atm		= 0.0717 k mol/hr	1 k mol/h	= 13.943 scfm
1 scfm at 70°F, 1 atm		= 0.0704 k mol/hr	1 k mol/h	= 14.212 scfm
1 scf at 60°F, 1 atm		= 0.02679 m³ at 0°C, 1 atm	1 m³ at 0°C, 1 atm	= 37.3258 scf at 60°F, 1 atm
1 scf at 70°F, 1 atm		= 0.02629 m³ at 0°C, 1 atm	1 m³ at 0°C, 1 atm	= 38.044 scf 70°F, 1 atm

Heat Transfer Coefficient

Units	W/m²K	kcal/m² hr°C	Btu/ft² hr°F
1 W/m² K	1	0.86042	0.17611
1 kcal/m² hr°C	1.16222	1	0.210468
1 Btu/ft² hr°F	5.67826	4.88570	1

Heat Flux

	W/m²K	kcal/m² hr°C	Btu/ft² hr°F
1 W/m²	1	0.86042	0.31700
1 kcal/m² hr°C	1.16222	1	0.36842
1 Btu/ft² hr°F	3.15459	2.71428	1

Length

U.S.	Sub-division	SI	SI	U.S.
1 inch		= 25.4 mm	1 cm	= 0.3937 in.
1 foot	1 ft = 12 in.	= 0.3048 m	1 m	= 1.09361 yd
1 yard	1 yd = 3 ft	= 0.9144 m	1 m	= 3.2808 ft
1 mile	1 mile = 1760 yd	= 1.609 km	1 km	= 0.62137 mile

Mass

U.S.	Sub-division	SI	SI	U.S.
1 pound	1 lb = 16 oz	= 0.4536 kg	1 kg	= 2.205 lb
1 short ton	1 st = 2,000 lb	= 0.907.185	1 t	= 1.1023 st
1 long ton	1 lt = 2,240 lb	= 1016.047 kg	1 t	= 0.90842 lt

Table A-12 Continued
Engineering Conversion Factors

Power			
Units	W	kW	HP
1 W = 1 J/S	1	10^{-3}	1.34102×10^{-3}
1 kW	10^3	1	1.34102
1 HP = 550 ft · lb/s	745.7	0.7457	1

Pressure				
Units	Pa	bar	atm	Psi
1 Pa = 1 N/m²	1	10^{-5}	0.98692×10^{-5}	1.4504×10^{-4}
1 bar = 10^5 Pa	10^5	1	9.98692	14.504
1 atm = 760 Torr	1.01325×10^5	1.01325	1	14.696
1 Psi = 1 lbf/in.²	6894.8	0.68948×10^{-1}	0.68046×10^{-1}	1

Specific Heat			
Units	J/kg K	kcal/kg °C	Btu/lb °F
1 J/kg K	1	2.39006×10^{-4}	2.39006×10^{-4}
1 kcal/kg °C	4.184×10^3	1	1
1 Btu/lb °F	4.184×10^3	1	1

Thermal Conductivity			
Units	W/m K	kcal/m hr°C	Btu/ft hr°F
1 W/m K	1	0.86042	0.57779
1 kcal/m hr°C	1.16222	1	0.67152
1 Btu/ft hr°F	1.73073	1.48916	1

Volume				
U.S.	Sub-division	SI	SI	U.S.
1 cubic inch		= 16.387 cm³	1 cm³	= 0.610 in.³
1 cubic foot	1 cf	= 28.317 dm³	1 dm³	= 0.0353 ft³
1 cubic yard	1 cy	= 0.76467 m³	1 m³	= 1.308 yd³
1 U.S. gallon	1 USG	= 3.785 dm³	1 m³	= 0.2642 U.S. gal.

  * According to National Bureau of Standards
 ** According to International Table

Table A-13
Conversion Factors—Alphabetic Listing

Multiply	By	To Obtain
atm	29.92	in. Hg
	33.90	ft water
	101.3	kPa
	14.70	lbf/sq in.
bbl	42.0	U.S. gal
Btu	1.055	joules
	777.5	ft-lb
	3.92×10^4	hp-hr

Table A-13 Continued
Conversion Factors—Alphabetic Listing

Multiply	By	To Obtain
Btu/min.	12.96	ft lbf/sec
	0.02356	hp
	0.01757	kW
	17.57	W
cm	0.3937	in.
	0.01	m
	10.0	mm
cm/sec	1.969	ft/min.
	0.03281	ft/sec
	0.036	km/hr
	0.6	m/min.
	0.02237	mi/hr
	3.728×10^{-4}	mi/min.
cu cm	3.531×10^{-5}	cu ft
	6.102×10^{-2}	cu in.
	10^{-6}	cu m
	1.308×10^{-6}	cu yd
cu ft	2.832×10^{4}	cm^3
	1,728	cu in.
	0.02832	cu m
	0.03704	cu yd
cu ft/min.	472.0	cm^3/sec
	62.43	lb of water/min.
cu in.	16.39	cu cm
	5.787×10^{-4}	cu ft
	1.639×10^{-5}	cu m
	2.143×10^{-5}	cu yd
cu m	10^{4}	cu cm
	35.312	cu ft
	61.023	cu in.
	1.308	cu yd
cu yd	7.7646×10^{5}	cu cm
	27	cu ft
	46.656	cu in.
	0.7646	cu m
cu yd/min.	0.45	cu ft/sec
degrees (angles)	60	minutes
	0.01745	radians
	3,600	seconds
degrees/sec	0.01745	radians/sec
	0.002778	revolutions/sec
drams	27.34375	grains
	0.0625	oz
	1.771845	grams
fathoms	6	ft
ft	30.48	cm
	12	in.
	0.3048	m
	0.333	yd

Table A-13 Continued
Conversion Factors—Alphabetic Listing

Multiply	By	To Obtain
ft water	0.02950	atm
	0.8826	in. Hg
	2.99	kPa
	62.43	lbf/sq ft
	0.4335	lbf/sq ft
ft/min.	0.5080	cm/sec
	0.01667	ft-lbf/sec
	0.01829	km/hr
	0.3048	m/min.
	0.01136	mi/hr
ft/sec/sec	30.48	cm/sec/sec
	0.3048	m/sec/sec
ft-lbf	1.286×10^{-3}	Btu
	5.050×10^{-7}	hp-hr
	1.36	joules
ft-lbf/min.	1.286×10^{-3}	Btu/min.
	0.01667	ft-lbf/sec
	3.030×10^{-5}	hp
	1.36	joules/min.
	2.260×10^{-5}	kW
ft-lbf/sec	7.717×10^{-2}	Btu/min.
	1.818×10^{-3}	hp
	2.267×10^{-2}	joules/min.
	1.356×10^{-3}	kW
gal (Imp)	0.1606	cu ft
	277	cu in.
	5.94×10^{-3}	cu yd
	4.546	l
	8	pints (liq.)
	4	quarts (liq.)
	1.20095	U.S. gal
gal (U.S.)	0.83267	imp gal
gal (Imp) water	10.02	lb water
gal (Imp)/min.	2.676×10^{-3}	cu ft/sec
	0.07577	l/sec
	9.6326	cu ft/hr
grains (troy)	1	grains (avoir)
	0.06480	grams
	0.04167	pennyweights (troy)
	2.0833×10^{-3}	oz (troy)
grains/gal (U.S.)	17.118	ppm
grains/gal (Imp.)	14.286	ppm
grams	980.7	dynes
	15.43	grains
	10^{-3}	kg
	10^{3}	mg
	0.03527	oz
	0.03215	oz (troy)
	2.205×10^{-3}	lb

Table A-13 Continued
Conversion Factors—Alphabetic Listing

Multiply	By	To Obtain
grams/cm	5.600×10^{-3}	lb/in.
grams/cm³	62.43	lb/cu ft
	0.03613	lb/cu in.
grams/l	48.642	grains/gal
	8.345	lb/1000 gal
	0.062437	lb/cu ft
	1,000	ppm
hectoliters (hl)	100	l
horsepower (hp)	42.44	Btu/min.
	33.00×10^3	ft-lbf/min.
	550	ft-lbf/sec
	0.7457	kW
	745.7	W
hp (boiler)	33.479×10^3	Btu/hr
	9.803	kW
hp-hr	2,547	Btu
	1.98×10^6	ft-lbf
	2.685	megajoules
	0.7457	kWh
in.	2.540	cm
in. Hg	0.03342	atm
	1.133	ft water
	3.38	kPa
	70.73	lbf/sq ft
	0.4912	lbf/sq ft
in. water	0.002458	atm
	0.07355	in. Hg
	249	pascals
	0.5781	oz/sq in.
	5.202	lbf/sq ft
	0.036133	lbf/sq ft
kg	980.665×10^3	dynes
	2.205	lb
	1.102×10^{-3}	tons (short)
	10^3	grams
kg/m	0.6720	lb/ft
kg/sq mm	10^6	kg/sq m
kl	10^3	l
km	10^5	cm
	3,281	ft
	10^3	m
	0.6214	mi
	1,094	yd
km/hr	27.78	cm/sec
	54.68	ft/min.
	0.9113	ft/sec
	0.5396	knots
	16.67	m/min.
	0.6214	mi/hr

Table A-13 Continued
Conversion Factors—Alphabetic Listing

Multiply	By	To Obtain
kW	56.92	Btu/min.
	4.425×10^4	ft-lbf/min.
	737.6	ft-lbf/min.
	1.341	hp
	14.34	kcal/min.
	10^3	W
kWh	3,415	Btu
	2.65×10^6	ft-lbf
	1.341	hp-hr
	3.6	MJ
liter (l)	10^3	cu cm
	0.03531	cu ft
	61.03	cu in.
	10^{-3}	cu m
	1.308×10^{-3}	cu yd
	0.2200	imp gal
	1.759	pints (imp.)
	0.88	quarts (imp.)
l/min.	5.886×10^{-4}	cu ft/sec
	3.666×10^{-3}	imp gal/sec
meter (m)	100	cm
	3.281	ft
	39.37	in.
	10^{-3}	km
	10^3	mm
	1.094	yd
m/min.	1.667	cm/sec
	3.281	ft/min.
	0.05468	ft/sec
	0.06	km/hr
	0.03728	mi/hr
m/sec	196.8	ft/min.
	3.281	ft/sec
	3.6	km/hr
	0.06	km/min.
	2.237	mi/hr
	0.03728	mi/min.
micron	10^{-6}	m
mi	1.609×10^5	cm
	5,280	ft
	1.609	km
	1,760	yd
mi/hr	44.70	cm/sec
	88	ft/min.
	1.467	ft/sec
	1.609	km/hr
	0.8684	knots
	26.82	m/min.

Table A-13 Continued
Conversion Factors—Alphabetic Listing

Multiply	By	To Obtain
mi/min.	2,682	cm/sec
	88	ft/sec
	1.609	km/min.
	60	mi/hr
mg	10^{-3}	grams
ml	10^{-3}	l
mm	0.1	cm
	0.03937	in.
mg/l	1	ppm
million gal (imp)/day	1.54723	cu ft/sec
miner's in.	1.5	cu ft/min.
minutes (angle)	2.909×10^4	radians
oz	16	drams
	437.5	grains
	0.0625	lb
	28.349527	grams
	0.9115	oz (troy)
	2.790×10^{-5}	tons (long)
	2.835×10^{-5}	tons (metric)
oz (troy)	480	grains
	20	pennyweights (troy)
	0.08333	lb (troy)
	31.103481	grams
	1.09714	oz (avoir)
oz (fluid)	28.41	ml
parts/million (ppm)	0.0584	grains/gal (U.S.)
	0.07016	grains/gal (Imp.)
	8.345	lb/million gal
pennyweights (troy)	24	grains
	1.55517	grams
	0.05	oz (troy)
	4.1667×10^3	lb (troy)
lb	16	oz
	256	drams
	7,000	grains
	0.0005	tons (short)
	453.5924	grams
	1.21528	lb (troy)
lb (troy)	5,760	grains
	240	pennyweights (troy)
	12	oz (troy)
	373.24177	grams
	0.822857	lb (avoir)
	13.1657	oz (avoir)
	3.6735×10^{-4}	tons (long)
	4.1143×10^{-4}	tons (short)
	3.7324×10^{-4}	tons (metric)
lb water	0.01602	cu ft
	27.68	cu in.

Table A-13 Continued
Conversion Factors—Alphabetic Listing

Multiply	By	To Obtain
lb water	0.0998	gal (imp.)
lb water/min.	2.670×10^{-4}	cu ft/sec
lb/cu ft	0.01602	grams/cu cm
	10.02	kg/m^3
	5.787×10^{-4}	lb/cu in.
lb/cu in.	27.68	grams/cu cm
	2.768×10^4	kg/m^3
	1,728	lb/cu ft
lb/ft	1.488	kg/m
lb/in.	178.6	grams/cm
lbf/sq ft	0.01602	ft water
	4.882×10^{-1}	g/cm^2
	6.945×10^{-3}	lbf/sq in.
lbf/sq ft	0.06804	atm
	2.307	ft water
	2.036	in. Hg
	0.0703	kgf/m^3
quarts U.S. (dry)	67.20	cu in.
temperature		
(°C) + 273	1	abs. temp. (K)
(°F) + 460	1	abs. temp. (R)
(°F) − 32	5/9	temp. (°C)
tons (long)	1,016	kg
	2,240	lb
	1.120	tons (short)
tons (metric)	10^3	kg
	2,205	lb
tons (short)	2,000	lb
	32,000	oz
	907.18486	kg
	2,430.56	lb (troy)
	0.89287	tons (long)
	29,166.66	oz (troy)
	0.90718	tons (metric)
watts (W)	0.05692	Btu/min
	44.26	ft lbf/min
	0.7376	ft lbf/sec
	1.341×10^{-3}	hp
W-hr	3.415	Btu
	2,655	ft-lbf
	1.341×10^{-3}	hp-hr
	3,600	joules
	10^{-3}	kWh

Note: Liquid units, unless specified otherwise, refer to imperial units (U.K.).

ABBREVIATIONS

AGA	–	American Gas Association
ANSI	–	American National Standards Institute
API	–	American Petroleum Institute
BTU	–	British Thermal Unit
CGI	–	combustible gas indicator
CI	–	cast iron
CNG	–	compressed natural gas
CP	–	cathodic protection
CPM	–	critical path method
CSA	–	Canadian Standards Association
CTS	–	copper tube size
CW	–	continuous weld
DC	–	direct current
DN	–	diameter nominal
DOT	–	Department of Transportation (U.S. Federal)
DP	–	design pressure
DR	–	dimension ration D/T (see also SDR)
EPC	–	engineering, procurement, construction
ERW	–	electric resistance weld
FS	–	forged steel
HP	–	high pressure
ID	–	inside diameter
IGT	–	Institute of Gas Technology, Chicago
IP	–	intermediate pressure
ISO	–	International Standards Organization
J	–	joule, a unit of energy
kPa	–	kilopascal
LP	–	low pressure
M&S	–	mains and services
mA	–	milliampere
MAOP	–	maximum allowable operating pressure
MCFH	–	1,000 cubic feet per hour (flow rate)
MI	–	malleable iron
MOP	–	maximum operating pressure
MPa	–	megapascal
MSS	–	Manufacturers' Standardization Society
NPS	–	nominal pipe size

OD	–	outside diameter
O&M	–	operation and maintenance
OSHA	–	Occupational Safety and Health Administration
PC	–	personal computer
PE	–	polyethylene
P.E.	–	professional engineer
PN	–	pressure nominal
PUB	–	Public Utility Board
PSI	–	pounds per square inch
PVC	–	polyvinyl chloride
Q	–	gas flow rate
R	–	See DR. R is also used as constant in gas flow equations
SCFH	–	standard cubic feet per hour (flow rate)
SDR	–	standard dimension ratio = D/t where D is outside diameter and t is pipe wall thickness. Commonly used dimension ratios (DR) are called SDR.
SI	–	system international (metric)
STD	–	standard
SST	–	stainless steel tubing
TLV	–	threshold limit value
TO	–	turn off
USG	–	United States gallon
wc	–	water column
WO	–	work order
WOG	–	water, oil, and gas (cold rating)
WT	–	wall thickness
XH	–	extra heavy
X^n	–	cross section
YS	–	yield strength
Z	–	gas supercomressibility factor
>	–	greater than
<	–	less than
≥	–	equal or greater than
≤	–	equal or less than
/	–	per, and
×	–	multiplied by
:	–	ratio
Δ	–	difference in
π	–	22/7
η	–	eta, efficiency factor

REFERENCES

1. *Canadian Metric Practice Guide* (Publication CAN/CSA-Z234.1-89), Canadian Standards Association, Rexdale (Toronto) Ontario, 1989. Reprinted with permission.
2. *The International System of Units (SI)* (Publication CAN/CSA-Z234.2-79), Canadian Standards Association, Rexdale (Toronto) Ontario, 1980. Reprinted with permission.
3. *Standard for Metric Practice* (Publication ASTM E 380-79), The American Society for Testing and Materials, Philadelphia, PA, 1980. Reprinted with permission.

Index